Sami Missaoui
Romdhane Ben Slama
Zied Driss

RECOVERY OF WASTE HEAT FROM REFRIGERATION MACHINES

Sami Missaoui
Romdhane Ben Slama
Zied Driss

RECOVERY OF WASTE HEAT FROM REFRIGERATION MACHINES

WATER HEATING AND DESALINATION

ScienciaScripts

Imprint

Cover image: www.ingimage.com

This book is a translation from the original published under ISBN 978-3-8416-3300-2.

Publisher:
Sciencia Scripts
is a trademark of
Dodo Books Indian Ocean Ltd. and OmniScriptum S.R.L publishing group

120 High Road, East Finchley, London, N2 9ED, United Kingdom
Str. Armeneasca 28/1, office 1, Chisinau MD-2012, Republic of Moldova, Europe
Managing Directors: Ieva Konstantinova, Victoria Ursu
info@omniscriptum.com

Printed at: see last page
ISBN: 978-620-3-49131-9

Contents

General introduction

Refrigeration machines such as fridges, freezers and air conditioners are now part of our daily lives. Through their various applications, they ensure a better quality of life. They can be used to freeze and preserve food products, or to cool premises such as supermarkets, flats, shops, cinemas and sports complexes. These refrigeration machines have cycles that extract heat from the medium to be cooled at a low temperature, releasing it to the outside environment at a higher temperature. This vapour-compression refrigeration cycle is a resounding success thanks to its reliability. This refrigeration equipment is designed to produce cold in all areas requiring cooling and/or freezing. Cooling is provided by a heat exchanger called the evaporator, while on the other side there is another heat exchanger called the condenser, which supplies a large amount of heat energy to the hot source at a higher temperature. This waste heat from the condenser will be lost to the atmosphere without being used. So these refrigeration machines produce two different types of heat energy, only one of which is used and the other, produced by the condenser, is released into the atmosphere, which in turn contributes to environmental degradation. Using this source of free energy lost in the environment is capable of satisfying heating and desalination needs at all temperatures while maintaining good thermodynamic performance. For this reason, why not make the most of the waste heat supplied by the condenser to heat domestic hot water or for desalination?

Domestic hot water heating and seawater desalination processes require high levels of energy consumption and therefore have a major impact on the environment. The advantage of using waste heat is that it has an influence on both economic factors (changes in energy costs) and environmental factors (reduction in greenhouse gas emissions). Improving energy efficiency makes it possible to reduce the energy factor on the market and therefore reduce the environmental impact. The latter is considerably reduced by the reduction in CO_2 emissions, which can account for a significant percentage of boiler emissions.

Currently, in industry, the application of this type of waste heat recovery is low, but given the progressive increase in energy costs and environmental problems, this number can only grow.

The aim of this work is study the heat released by the condensers of refrigeration machines for water heating and desalination in order to optimise the operating conditions of these machines.

This is done through a theoretical and experimental study based on the heat exchange phenomena between the refrigerant inside the tube and the water outside, which govern the heating and desalination of this fluid.

This report is divided into four chapters.

The first chapter gives a general description of compression refrigeration machines and heat pumps, as well as techniques for domestic hot water and even desalination of this substance using waste heat from the condenser.

An experimental study of a domestic refrigerator coupled with a water heater and a air conditioner that distils brackish water will be subject of the second chapter. A detailed description of the prototypes, their characteristics and geometric parameters are then presented.

The third chapter will be devoted to modelling the operation and coupling of the domestic refrigerator to the water heater.

The fourth and final chapter deals with the results obtained and their discussion.

Finally, the report concludes with an outlook on the study.

CHAPTER 1

Bibliographical study

1 Introduction

This chapter begins with a few general notions about compression refrigeration machines and heat pumps, and a few basic elements about the principle of thermodynamics. We then describe domestic water heating technology, presenting the different heating methods used by researchers in this field and their results. Next, we present some information on the various seawater desalination technologies, in particular membrane desalination and desalination by phase-change distillation.

2 General information on refrigeration machines

Refrigeration machines such as refrigerators, freezers and air conditioners have two heat exchangers: the evaporator to create cold by evaporating freon, and the condenser to transfer heat to the outside environment by condensing freon. These refrigeration machines produce heat and cold simultaneously. The main purpose of the refrigeration machine is to produce cold for domestic, commercial and industrial needs, which requires the use of a device capable of extracting heat from the medium to be cooled and transferring it to an external medium.

2.1 Operating principle

A refrigeration plant transfers heat energy from a low-temperature medium to be cooled to an external medium at a higher temperature. Freon is commonly used for this transfer. This refrigerant circulates in the refrigeration system. The vapour compression refrigeration machine is made up of four main components: the compressor, condenser, expansion valve and evaporator, as shown in Figure 1.1.

Freon follows a closed four-phase cycle through the circuit made up of the main components:

- Compression of freon (gaseous state).
- Condensation of freon (gaseous then liquid state).
- Freon expansion.
- Vaporisation of the liquid fluid (cold production).

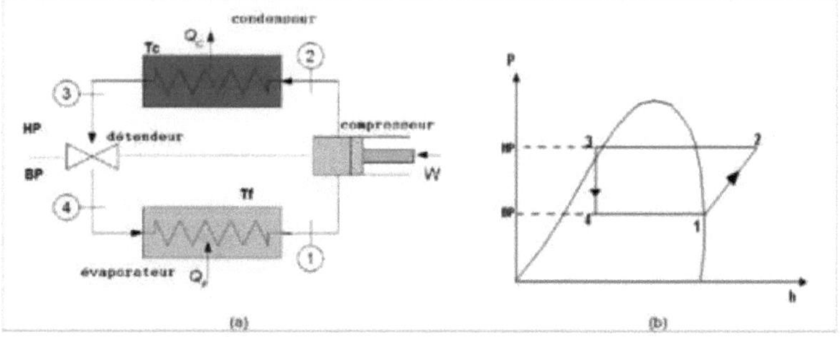

Figure 1. 1 (a) Schematic diagram of the refrigeration machine, (b) Ideal cycle of the machine **[1].**

2.2 Components of a refrigeration machine

2.2.1 Compressors

The compressor delivers the superheated freon (refrigerant) at high temperature and high pressure to a heat exchanger called the condenser, which is located in an external environment at a lower temperature than the freon.

2.2.2 Condenser

The condenser is a heat exchanger that receives the refrigerant compressed by the compressor to high pressure and high temperature. The hot freon from the compressor supplies its heat to the outside environment. Heat energy is therefore transferred from the freon to the outside environment. The refrigerant cools, resulting in condensation. At the outlet of the exchanger (condenser), the result is freon in a cooled liquid state at high pressure.

2.2.3 Pressure reducer

The freon in its liquid state, cooled to high pressure, is then fed into the expansion valve, which expands adiabatically and without any work being supplied or consumed. The enthalpy at the inlet and outlet of the expansion valve is therefore the same. At the outlet, the freon is a liquid-vapour mixture at low temperature and low pressure. This then feeds the evaporator.

2.2.4 Evaporator

The evaporator is placed in the medium to be cooled. Inside the evaporator, the refrigerant is a liquid-vapour mixture at low pressure and low temperature. The medium to be cooled emits heat at a higher temperature than the freon, so heat energy is transferred from the medium to be cooled to the freon. The refrigerant heats up, causing it to evaporate. The result, at the evaporator outlet, is freon in a superheated vapour state at low pressure.

2.3 Field of application

Refrigeration machines are used in a number of areas. By producing cold, they meet the needs users in different locations. They can be used to freeze and preserve food products. Refrigeration equipment makes an essential contribution to socio-economic development. They have played a vital role in feeding the population by preserving foodstuffs during transport, distribution and presentation for sale. They have also been used in the health sector to preserve vaccines and cool operating theatres, not forgetting their use in housing, such as flats, hotels and houses, for refrigeration, freezing and air conditioning. These refrigeration machines can also be found in other areas, such as cooling workplaces in geographically hot and humid areas, and in large retail outlets and cinemas. Refrigeration is also needed in the electronics manufacturing industry, for example in the creation of microprocessors and air treatment in computer rooms.

2.4 Basic elements of thermodynamics

2.4.1 Mollier diagram

The enthalpy diagram shown in Figure 1.2 can be used to plot the cycle of a refrigeration machine and to determine the physical quantities of a refrigerant in a refrigeration plant, i.e. pressure, temperature, enthalpy, entropy and mass volume. This diagram can also be used to calculate the heat output of the two heat exchangers, the work done by the compressor and the state of the fluid at different points.

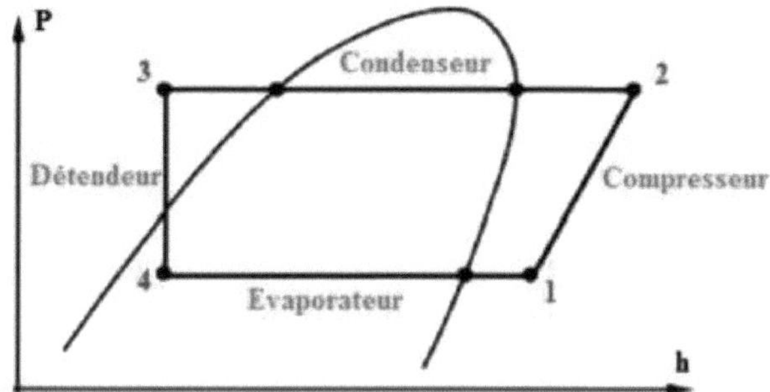

Figure 1. 2. Enthalpy diagram of the refrigeration cycle **[2].**

2.4.2 Energy conservation for the refrigeration machine

The conservation of total energy in a closed system is the first principle of thermodynamics, formulated by Julius Robert Von Mayer in 1842 and verified by experimental work in 1843 by James Prescott Joule. This principle can be written as :

$$\Delta E = \Sigma Q + W \tag{1-1}$$

Where E is the total energy exchanged by the system, $\Sigma\ Q$ is the sum of the heat quantities of the two exchangers and W is the mechanical work of the compressor. These energies are defined as equivalent forms of energy as shown in Figure 1.3.

This diagram illustrates the principle of the refrigeration machine. Using the mechanical work W of the compressor, the refrigeration cycle transfers heat from a cold source Qf at a temperature Tf to a hot source Qc at a higher temperature Tc.

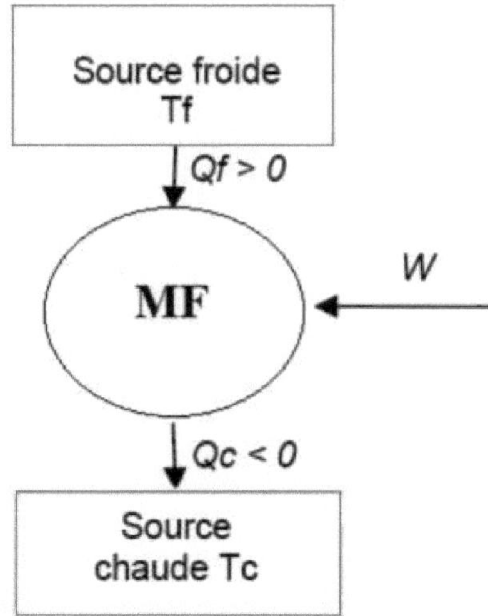

Figure 1. 3: Principle of the refrigeration machine

2.4.3 Condenser cooling capacity

The heating capacity supplied by the condenser can be identified according equation (1-2) from the point of view of the refrigerant. From the coordinates of the points in the enthalpy diagram from the point of view of the hot-source fluid, we can write :

$$\dot{Q}_{c-r} = \dot{m}_r \cdot (h_3 - h_2) \qquad (1\text{-}2)$$

Where mir is the freon flow rate and h_2 and h_3 are the enthalpies at the condenser inlet and outlet respectively.

Or

$$\dot{Q}_{c-sc} = \dot{m}_{sc} \cdot Cp_{sc} \cdot \Delta T_{sc} \qquad (1\text{-}3)$$

Where $Q_{s\text{-}cs}$ is the flow rate of the cooling medium, Cp_{sc} is se specific heat and ΔT'^. is the difference in temperature between the inlet and outlet.

2.4.4 Evaporator cooling capacity

In the same way as for the condenser, the cooling capacity of the evaporator can be identified using the following formula:

$$\dot{Q}_{f-r} = \dot{m}_r \cdot (h_1 - h_4) \qquad (1\text{-}4)$$

Where mir is the freon flow rate and h1 and h_4 are the inlet and outlet enthalpies at the evaporator respectively.

For the source fluid, the formula can be written as follows:

$$\dot{Q}_{f-sf} = \dot{m}_{sf} \cdot Cp_{sf} \cdot \Delta T_{sf} \qquad (1\text{-}5)$$

Where $\dot{m}_{sf}$ is the flow rate of the cooling medium, Cp_{sf} its heat capacity and ΔT_{sf} is the temperature difference between the inlet and outlet.

2.4.5 Calculating the electrical power absorbed by the compressor

According to the principle of thermodynamics, the power absorbed by compression is defined by the product of the freon mass flow rate m_r and the difference in enthalpy between the compressor inlet and outlet:

$$\dot{W} = \dot{m}_r \cdot (h_2 - h_1) \qquad (1\text{-}6)$$

The Freon mass flow rate is calculated using the following formula:

$$\dot{m}_r = \rho_r \cdot \eta_{vol} \cdot V_b \qquad (1\text{-}7)$$

Where ρ_r is the density of the freon, η_{vol} is the volumetric efficiency of the compressor and V_b is the swept volume.

The electrical power absorbed by the compressor is determined by the following formula :

$$P_{élec} = \frac{\dot{W}}{\eta_{él} \cdot \eta_m} \qquad (1\text{-}8)$$

Where W is the compressor power, nel is the electrical efficiency and nm is the mechanical efficiency of the compressor.

2.4.6 Performance evaluation

Several coefficients can be used to evaluate the energy performance of a compression refrigeration machine. In general, the performance of a system is defined by the ratio of the useful power Pu produced by the machine in the form of heat by the refrigerant to the power absorbed Pa by the compression in the form of mechanical or electrical work, as indicated by the following formula:

$$COP = \frac{Pu}{Pa} \quad (1\text{-}9)$$

Referring to the evaporator of the refrigeration machine, the coefficient of performance is written as follows

$$COP_{froid} = \frac{\dot{Q}_f}{P_{él.a}} \quad (1\text{-}10)$$

Referring to the condenser of a heat pump alone, we can write :

$$COP = \frac{\dot{Q}_c}{P_{él.\ a}} \quad (1\text{-}11)$$

Referring to both the evaporator and the condenser, we can write :

$$COP = \frac{\dot{Q}_c+\dot{Q}_f}{P_{él.\ a}} \quad (1\text{-}12)$$

2.4.7 Heat transfer

At the boundary between two systems with two different temperature levels, heat is exchanged without any work being done. The size of the flow is proportional to the heat exchange surface and the temperature difference. The following exchange phenomena can be distinguished:

- Heat transfer without change of state (e.g. gas/gas, liquid/liquid).
- Heat transfer with a change of state of at least one of the two flows (for example evaporation, condensation).
- Combined transfer of heat and matter by convection and evaporation (e.g. cooling by water/air vaporisation).

To determine the heat transfer, the heat flow Q can be calculated using the following formula:

$$\dot{Q} = K \cdot A \cdot \Delta T_{me} \quad (1\text{-}13)$$

With :

$$\Delta T_{me} = \frac{\Delta T_e - \Delta T_s}{\ln(\Delta T_e/\Delta T_s)} \quad (1\text{-}14)$$

K: Heat transmission coefficient $W/m^2 K$.

A: Heat transmission surface m^2.

AT_{me} : Mean logarithmic difference in temperature K.

The average logarithmic temperature difference is also called the thermal length and is calculated from the effective average temperature difference in the exchanger: ATe: Difference in inlet temperature K.

ATs : Temperature difference at output K.

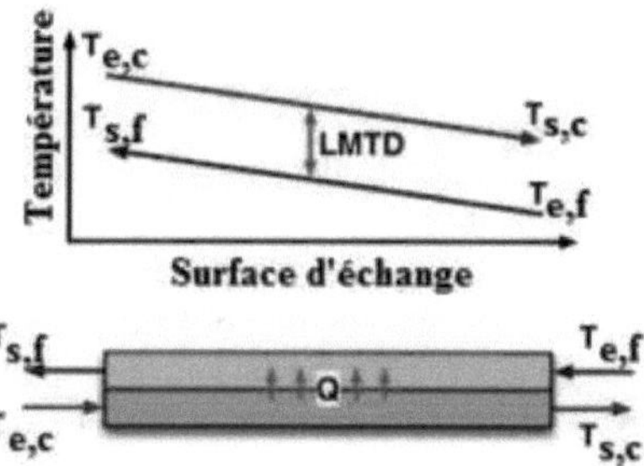

Figure 1. 4: Fluid temperature variation with countercurrent flow [3].

The heat transfer coefficient is not a characteristic dimension of the material, but is derived from many factors, such as the state of the fluid (vapour, gas, liquid), the properties of the fluid (density, specific heat value, viscosity, transmission capacity), the thermodynamic characteristics (pressure, temperature), the speed of the fluid and the dimensions and geometry of the heat exchanger.

It is impossible to establish a general formula for calculating the heat transmission coefficient, given all the interdependencies involved. In practice, this coefficient is defined for each case on the basis of experience or test values. Table 1.1 gives indicative values for various technical applications.

Table 1.1: Indicative values for heat transfer coefficients **[3]**.

Type	Heat flux (W/m²)
Water/water Tubular exchanger Double jacket exchanger	200 à1000 350à1400
Steam/water Tube exchanger (steam in the tube) Evaporator (steam around the tube)	350à1200 600à1500
Water/ gas Tubular exchanger Post-heating boiler (gas in the tube)	15 à 70 15 à 45
Gas/ gas Tubular exchanger Double jacket exchanger	6 à 35 2 à 35

2.4.8 High-temperature waste heat and heat requirements

Compression refrigeration machines produce heat and cold simultaneously using two heat exchangers. The first exchanger, the condenser, is in contact with the hot source, while the second exchanger, the evaporator, produces the cold. These machines have a refrigeration cycle in which a refrigerant circulates under pressure through a mechanical compressor. At the condenser, this refrigerant transfers heat energy to the outside environment at a very high temperature. On the other hand, on the evaporator side, the fluid receives a quantity of heat energy to produce cold for domestic or commercial use. By using just one heat exchanger, the evaporator, to meet our daily needs, the other heat exchanger, the condenser, loses a significant amount of heat to the atmosphere. These quantities of heat lost to the atmosphere are capable of satisfying heating needs at high temperatures (70°C to 80°C) while maintaining good thermodynamic performance. For this reason, why not make the most of this waste heat to improve the traditional heating technique, which consumes a lot of energy and produces gas emissions that are harmful to the environment.

3 Water heating

3.1 Production of domestic hot water

As the population has grown, so has the demand for energy, which has contributed to environmental pollution. In the Mediterranean region, the production of hot sanitary water is a major factor in residential energy consumption. Reducing the demand for water-heating energy conserves energy and protects the environment from gas emissions. Around 75% of the population electricity to heat water, 22% uses wood, oil or gas to heat water and 3% uses solar water heaters **[2]**. In Hong Kong, a large amount of domestic hot water is consumed in residential and commercial flats. With a low energy source, fossil fuel is the major means used to satisfy the need for hot water. With the deterioration of energy and environmental pollution, researchers have found a suitable solution to meet the need for water heating. This method is solar water heating, but it is insufficient in some regions where solar energy is scarce. For this reason, they proposed other heating techniques by coupling heat pumps and solar water heaters **[4]. Chen et al [5]** have shown that water heating technology is a major factor in energy consumption in the residential sector. This was achieved using different energy sources such as natural gas or electric heating. These different heating sources have a direct or indirect influence on fossil fuel energy consumption. The heat pump technology used to heat water is a feasible method of conserving energy and also provides the water with efficient energy to heat it. Energy conservation and environmental protection are two terms that need to be considered. The use of solar energy, geothermal energy, air or other renewable energies are heating techniques that have helped to reduce the problem of high energy consumption. On the other hand, the heat pump heating technique is an effective way of heating water and reducing the amount of energy consumed by other heating techniques **[6].** Comparing the amount of energy consumed by the electric water heater and that consumed by the heat pump to heat domestic hot water, we find that percentage of energy conserved is between 40 and 60% **[7].** As far as domestic water heating technology is concerned, several methods have developed to meet this demand.

3.2 Condenser heating technology for heat pumps

Dai et al [8] carried out a numerical study on a heat pump coupled to a hot water production unit as shown in Figure 1.5. The results show that the heat transfer coefficient and the performance of the coupled system are improved with the modification of the condenser geometry. Also, they found that the helical condenser with variable diameter shows better performance compared to the simple helical condenser with constant diameter.

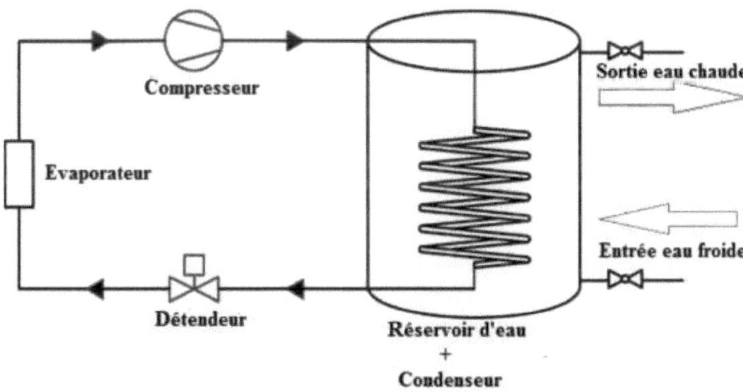

Figure 1. 5: Diagram of a heat pump coupled to a water heater **[8].**

Ye et al [9] presented a numerical study for a heat pump with helicoidal condenser wound on the outer wall of a hot water storage tank as shown in Figure 1.6. They found that the variable pitch helicoidal condenser showed better performance than the simple constant pitch helicoidal condenser.

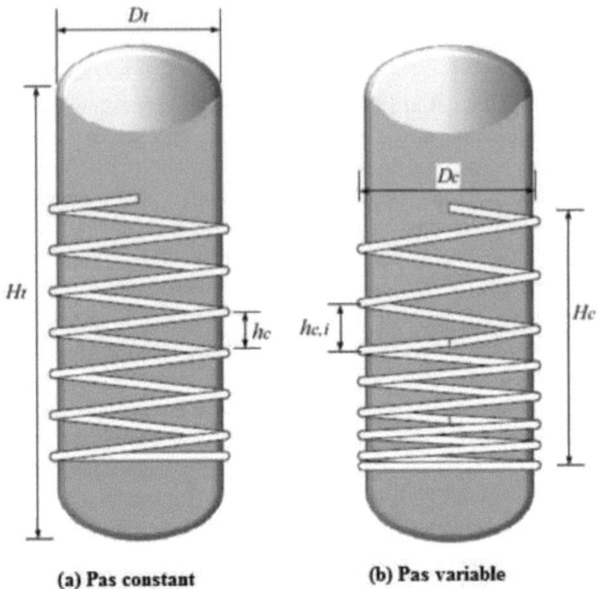

Figure 1. 6: Geometry of the condenser wound on the outer wall of the water tank **[9].**

Genkinger et al [10] studied the coupling of a heat pump with solar energy produce domestic hot water. They found that the reduction in energy consumption to meet residents' hot water needs is very significant. So they coupled a heat pump with a photovoltaic system to produce electricity using a solar panel to run the heat pump. The steam superheated in the condenser was used to heat domestic hot water. They showed that heating domestic water and producing energy using solar energy (photovoltaics) could change the outcome of the financial evaluation in the near future. The electricity produced by larger photovoltaic systems is already significantly cheaper than that produced by the smaller systems applied in this study. The technique of heating water through this system helps to reduce the percentage of energy lost. The results of this study are based on an assessment of the economic aspects.

Mahdi et al [11] conducted an experimental study on the melting behaviour of a phase change material (PCM) in an energy storage unit with a helically shaped condenser as shown in Figure 1.7, to investigate the effect of condenser geometry on the thermal performance of a coupled system. The results indicated that the melting rate in the cone-shaped condenser is 22% better than that in the helical-shaped condenser.

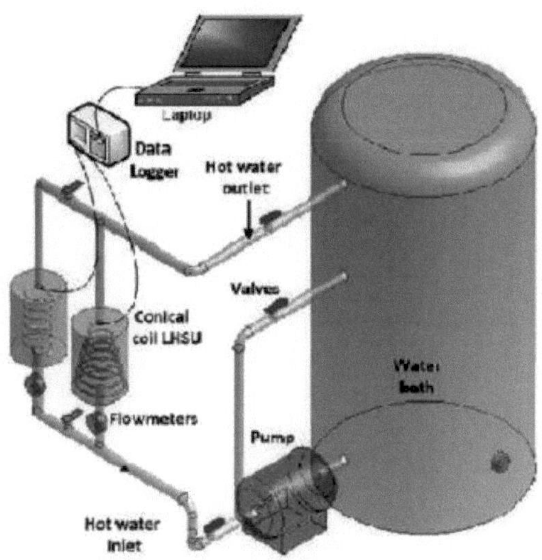

Figure 1.7: Heat pump coupled with a water heating unit **[11]**.

Skrivan [12] has used the heat of condensation from medium capacity chillers to heat domestic hot water, as shown in Figure 1.8. He showed that cooling installations offer favourable temperature conditions for heat recovery. He has also shown that in a farm producing milk, refrigeration is necessary to preserve the product. Hot water production is also essential for hygienic and technological purposes. So, as far as water heating technology is concerned, the only method used to satisfy daily needs is electrical energy. With the rising cost of electrical energy consumption and its impact on the environment, both of which are energy problems, this researcher has found a reliable method that will help to reduce electrical energy consumption. One way of achieving considerable savings is to use the heat of condensation from refrigeration units to produce hot water.

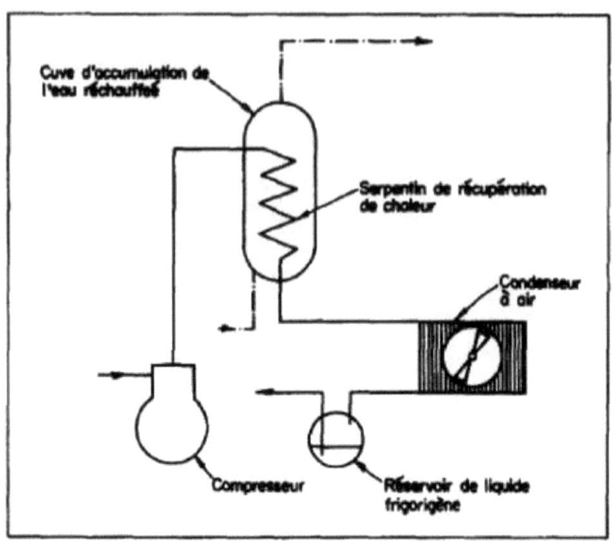

Figure 1. 8: Installation for heating water using the heat of condensation **[12].**

It has shown that using the waste heat from the refrigeration unit's condenser represents a clear energy saving. Some of the heat of compression is lost by radiation from the compressor, which slightly reduces the amount of heat available for recovery.

To optimise energy savings, the condensing temperature and therefore the temperature of the water heated should be adjusted according to the outside temperature and the characteristics of the expansion valve.

Carbonell et al [13] coupled a heat pump with a solar panel to produce heat in a cold climate and to prepare domestic hot water for daily use. In their work, they compared several different climates to check their influence on energy production for the solar panel, which in turn contributes to the production of warm air and the heating of domestic hot water using the heat pump's condenser.

Flora et al [14] studied a heat pump used for domestic water heating with different compressor speeds, as shown in Figure 1.9.

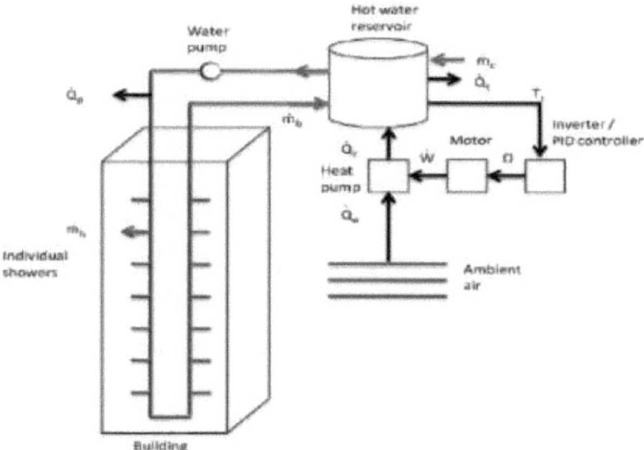

Figure 1. 9: Heat pump control technology for hot water distribution

[14]

Domestic hot water is used in very large quantities. To meet vital needs, most of the heating methods used electric. This is a very simple method with low installation costs. However, it is very expensive in energy terms. This study was carried out in a region of Brazil where most of the industrial companies consume around 80% of electrical energy. During the day, electrically heated showers are responsible for high consumption in the morning (from 5am to 9am) and especially in the early evening (from 6pm to 9pm). Given these problems, these researchers came up with an effective solution to solve them once and for all. They have shown that the advantage of this method is that it consumes less electricity than the traditional method.

The heat pump system is an effective way of reducing electrical energy consumption, leading to an approximate 73% reduction in total electricity consumption.

Li et al [15] carried out a study of the performance of solar assisted heat pump systems for hot water production in Hong Kong as shown in Figure 1.10.

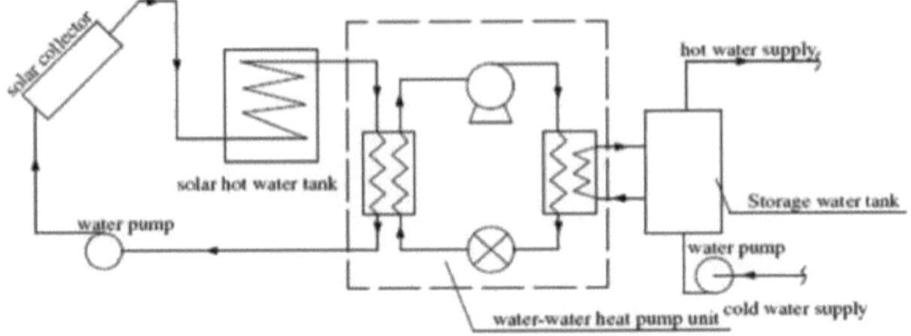

(a) Schéma de principe du système SA-WSHP

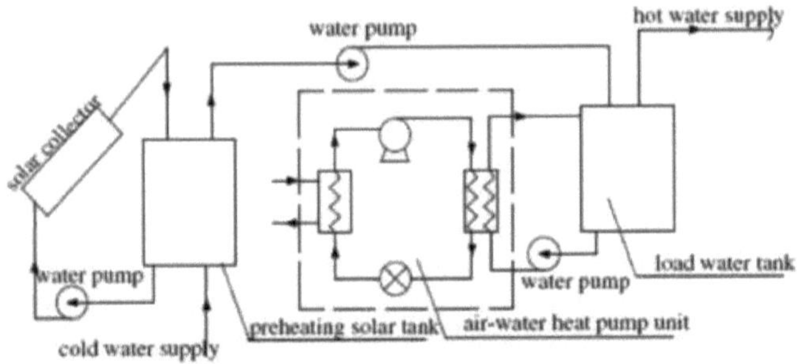

(b) Schéma de principe du système SA-ASHP

Figure 1. 10: Operating principle of the solar panel coupled with the heat pump **[15]**

In Hong Kong, a large quantity of hot water is used for residential and commercial purposes. The energy source used to produce hot sanitary water is fossil fuel. With the deterioration of energy and environmental pollution, solar water heating systems are helping to avoid these problems. They have also shown that on cold days or at night (when there is no solar radiation), the solar panel circuit does not work to heat the water, so the heat source recovered from the heat pump condenser starts to work. The operating principle of this system is the same as that of the conventional solar water heater, but its coupling with the heat pump leads to the production of auxiliary energy that contributes to the heating domestic hot water. They also showed that the flow rate of water from the tank to the condenser plays an important role in the system's performance, the final temperature of the water at the outlet and the total heating time. When the initial temperature of the water in the preheating tank is between 20 and 36°C, the flow rate varies from 4.7 m^3/h to 14.4 m^3/h. These results are in agreement with **Mei et al [16]** who studied the phenomenon of water heating by the heat pump condenser using natural convection.

Qin et al [17] carried out an experimental study on the technique of heating water using a domestic heat pump. They showed that heat pumps have an advantageous source of hot air, terms of efficiency, over natural gas boilers in terms of the need to heat domestic hot water. Measurements show that bubble diameter increases with a higher hot water flow rate. As the hot water flow rate increases, the water regime tends to be more turbulent.

Sun et al [7] studied the performance of the heat pump to heat water and also the heating technique by the combination of solar water heater and heat pump under a variable operating condition (air temperature, water temperature, solar radiation, intensity, ...) to make a comparison between the two systems as shown in Figure 1.11.

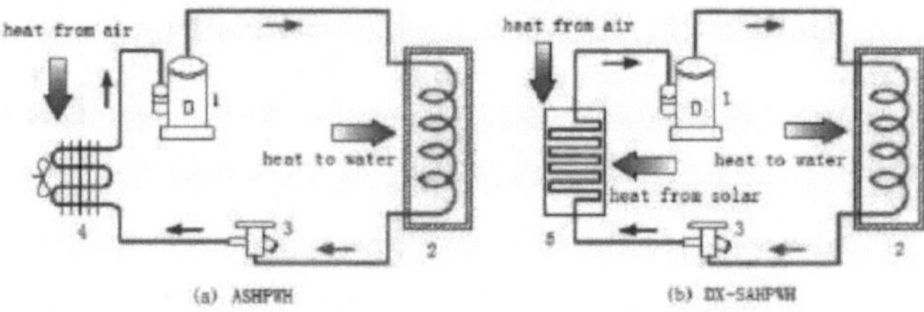

1-Evaporateur; 2-Réservoir d'eau; 3-Détendeur; 4-Evaporateur/ Ventilateur; 5-Evaporateur

Figure 1. 11. Block diagram of two systems: water heating using the heat pump's hot air source/solar water heater coupled to a heat pump **[7].**

A comparison between the electric resistance water heating system and the heat source supplied by the heat pump condenser shows a reduction in energy consumption of around 40 to 60%. The ambient temperature influences the performance of this system (heat pump hot air source for water heating "ASHPWH"). Low ambient temperature is an obstacle to the performance of this system. Water heating by the hot air source delivered by the heat pump condenser has better operating stability except in bad climates (low ambient temperature). They also showed that the coupled system (solar water heater with a heat pump) operates every day and in all weather conditions with greater profitability. They also showed that on a hot day, the COP of the solar water heating system coupled with a heat pump is higher than the other, which is defined as heating by a heat pump alone. On the other hand, during the night, the opposite is true, as shown in Figure 1.12. They have shown that on a clear day, the COP of the solar water heating system coupled with a heat pump is higher than that of the heat pump condenser water heating system. This difference is explained by the change in the evaporation temperature of the refrigerant carried out in the heat pump by the solar energy.

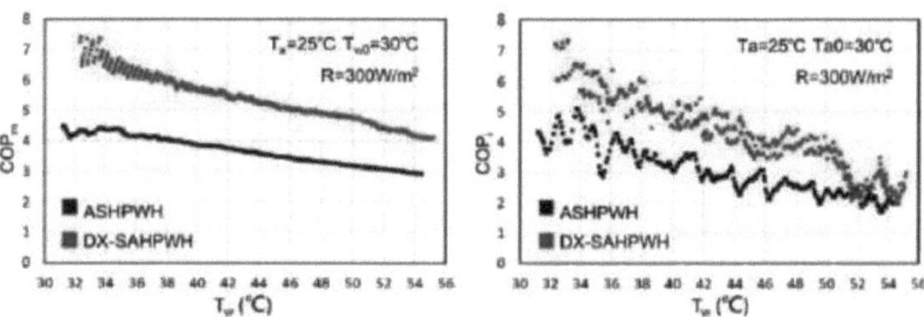

Figure 1. 12. COP comparison of two proposed systems **[7].**

Xinhui Zhao et al [6] carried out an experimental study on the heating performance of water in a tank a high-temperature heat source supplied by the heat pump condenser. They found that the temperature of the refrigerant (R410a) in the tank increases as the water temperature rises, which varies between 7 and 9°C.

The increase in water temperature is linked to the increase in energy supplied to the heat pump condenser. They have also shown that the condensation temperature is too high, increasing heat loss and reducing the COP.

When the water temperature is around 21 T, the heating capacity is 1870 W and if the water temperature increases by 1 T, the heating capacity decreases by 25 W; when the water temperature reaches 55 T, the heating capacity is 490 W, and if the water temperature increases by 1 °C, the heating capacity reduces to 80 W. As the temperature inside the water tank increases, the COP value is gradually reduced, as shown in Figure 1.13.

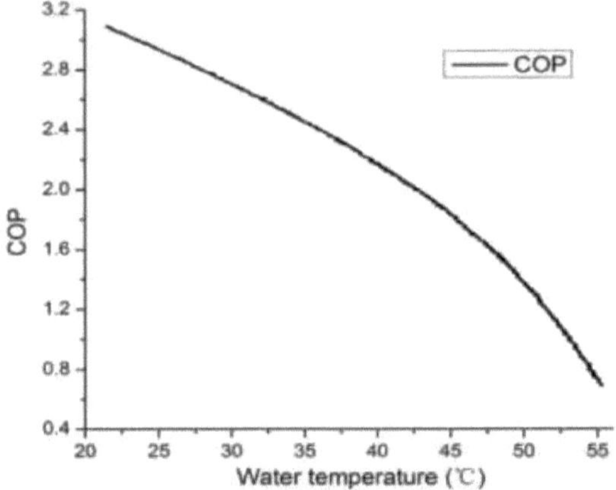

Figure 1. 13: Variation in COP as a function of water temperature **[6].**

Ben Slama [18] studied the principle of heating water through the refrigerator condenser. He showed that when this heating system is used, the temperature of the hot water reaches 60°C. In economic terms, heating water through the refrigerator condenser helps to protect the environment from CO2 emissions (environmental pollution) and reduce energy consumption.

Youssef et al [19] studied a coupled system: solar water heater with a heat pump to produce hot sanitary water. They reported that fossil-fuel methods of heating water produce co2 gases that have an impact on the environment. The solar water heating technique is also insufficient in cold (not stable) climates. For these problems, they looked for a method that could reduce energy consumption and protect the environment against the emission of gases and to satisfy the inhabitants' need for hot water in all conditions and in different climates. They found that this system is more cost-effective in all weather conditions (absence or presence of solar radiation).

The difference between COP on a hot day and on a cold day varies by around 6.1% and 14% respectively.

Tianji Liu et al [20] experimentally studied the principle of water heating by heat pump using solar energy. This system is based on the principle of heating water by the source of hot air supplied by the heat pump condenser and also by the solar water heater backed up by an electric resistor. This system has shown a COP around 82% at ambient temperature. The advantage of this system is that it heats domestic hot water when there is little or no solar energy (solar radiation).

Bakirci et al [21] experimentally studied the thermal performance of a solar-assisted heat pump system for domestic heating in a cold climate. They showed that the principle of heating by heat pump reduces the gas emissions produced by the other heating technique in residential buildings or commercial sectors. The reduction in CO2 emissions ranges from 15% to 77% when the heat pump is used for heating. They have shown that the heat pump method of heating is an important economic method compared with other methods of using fossil fuels (natural gas, alcohol, fuel oil, etc.). The property of the heat pump system represents an important contribution to reducing the high investment .

Ibrahim et al [2] studied the source of hot air supplied by the heat pump condenser for use in water heating. They showed that the reduction in gas emissions is significant for this system compared with other methods used, such as heating with natural gas. They also showed that climate differences influence the coefficients of performance of the system studied (heat pump).

Dott et al [22] evaluated a coupled system between a solar water heater and a heat pump for the production of hot water as shown in Figure 1.14. They showed that the heat pump is an important factor in producing sufficient heat to carry out the process of heating a space or domestic hot water. However, the only drawback is the high energy consumption due to the consumption of the compressor and the fan of this refrigeration machine.

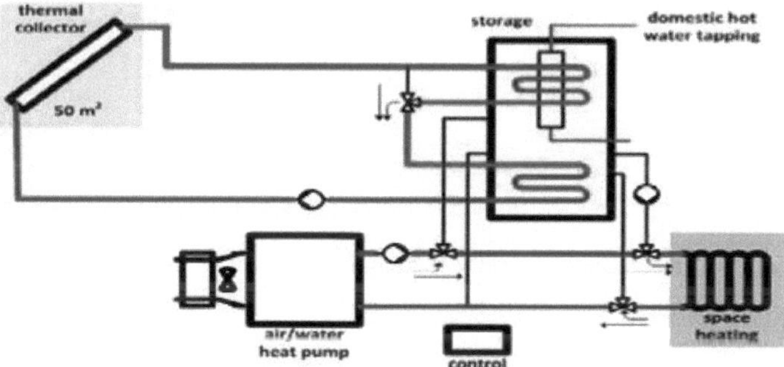

Figure 1. 14: Diagram of the coupled system (heat pump with solar panel) **[22].**

Eicher et al [23] studied the solar-assisted heat pump producing hot water. They used solar energy with the heat pump to improve the performance of this system. They showed that the COP decreased as the temperature at the condenser increased.

Ben Slama [24] studied a refrigerator used for heating (water and space heating) and also to reduce energy and carbon emissions. He showed that the refrigerator can contribute to water heating with the main function of producing cold. He observed that the temperature of hot water, heated by the refrigerator condenser, reaches 60°C. The condenser of a refrigerator immersed in a water tank represents a better source of heating and a reduction in energy consumption.

Jing-Wei Peng et al [25] compared the performance of the heat pump water heating system with different types of expansion valves. They found that the short orifice tube type expansion valve is better than the other capillary type for water heating technique. The capillary tube is used in the heat pump because of its low cost and simple manufacture.

3.3 Refrigerator condenser heating technology

Momin et al [26] determined the COP of a domestic refrigerator used to heat domestic water using the heat lost through its condenser. They found that the heat lost to the atmosphere by the refrigerator's condenser is very high compared with the ambient temperature. This waste heat is capable of satisfying domestic needs. For this reason, they used this waste heat to heat the water and improve the economic factor. They have shown that the maximum temperature obtained in the water tank reaches 60°C. The COP of the system using waste heat to heat the water is better than that obtained by condensing freon with air. They also found that power consumption was lower when water was used to condense the heat transfer fluid than when air was used. They also found that the heat supplied by the hot water heated by the refrigerator condenser can be used in other applications, such as space heating, domestic hot water, cleaning, washing or drying clothes.

Soni et al [27] studied a technique for using waste heat from a domestic refrigerator to heat water and air. They found that the technique of heating water using waste heat from the condenser is a free energy heating tool. The amount of waste heat reaches around 60°C and is capable of heating the water after 6 hours. This amount of energy is capable of meeting some or all of our domestic hot water needs for the whole day. This system enables us to energy and protect the environment from gas emissions. They have shown that the system used to recover waste heat has helped to improve the COP and also reduce energy consumption. The difference in water temperature at the condenser inlet and outlet around 10°C. On the other hand, the air temperature at the cabin boundary is around 46°C. They also concluded that this system can be used in a variety of places, such as hotels, hospitals, restaurants and for domestic purposes such as cleaning, washing, drying, etc.

Jadhav et al [28] used the waste heat from the refrigerator condenser to heat the water. They found that the amount of waste heat from the condenser is sufficient to meet the heating requirements of an enclosed space. They indicated that the refrigerator with a closed chamber and a water heating space are based on the same principle of the vapour compression cycle but with a small modification as shown in Figure 1.15.

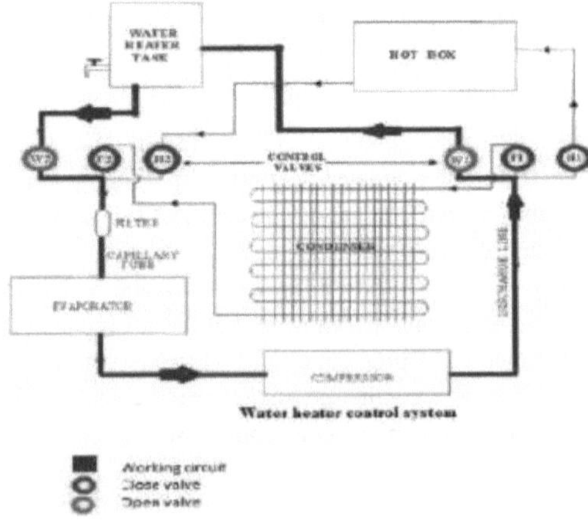

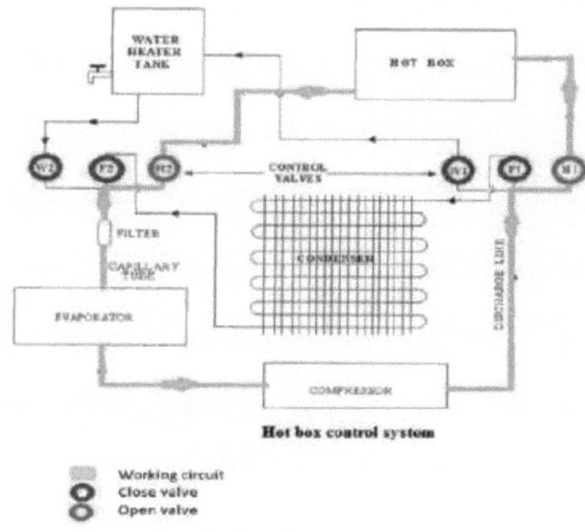

Figure 1. 15. Recovery of waste heat for heating **[28].**

They found that a refrigerator coupled with an enclosed hot water tank and a water heating space gave the best results. The temperature of the hot water in the tank varies between 55 and 60°C and the temperature level in the closed chamber is between 45 and 50°C. They have shown that this refrigeration machine, with its simultaneous production of heating and cooling, can be used in a wide range of applications, including hotels, domestic needs and industry. The simultaneous heating and cooling used in various locations is achieved by a process known as machine-multifunction.

Chaudhari et al [29] carried out an experimental study on a refrigerator to recover waste heat from its condenser. They showed that recovering this waste contributes to reducing energy and environmental pollution. They found that the amount of heat recovered by the refrigerator condenser varies between 375 and 407 W. The theoretical COP of the system without waste heat recovery around 1.88, but with waste heat recovery, the COP becomes equal to 2.53. The COP of the air-cooled condenser is 1.078. They also showed that the quantity of hot water available per hour is 7.2 litres at a temperature of 51°C and 24 L/h at a temperature of 41.7°C.

Sreejith et al [30] experimentally studied a domestic refrigerator using water and air to cool the condenser. They determined the performance of a refrigerator by comparing two sources of condenser cooling, namely air and water. They showed that the performance of a refrigerator is improved by using water to cool the condenser. Under all loading conditions, the water-cooled condenser contributes to a reduction in energy consumption of 8 to 11% compared with air-cooling technology. The quantities of hot water produced by a refrigerator's condenser cooling system are sufficient to meet domestic needs such as cleaning and washing.

Borkar et al [31] studied the use of waste heat from refrigerators. They found that the refrigerator condenser releases a quantity of heat energy into the atmosphere, which reduces the efficiency of this refrigeration machine. These researchers thought of using this waste heat for different applications to increase the efficiency of the refrigerator. The system studied is a domestic refrigerator that simultaneously produces heat and cold. The hot side, the condenser, heats the food and the cold side, the evaporator, cools the water, as shown in Figure 1.16.

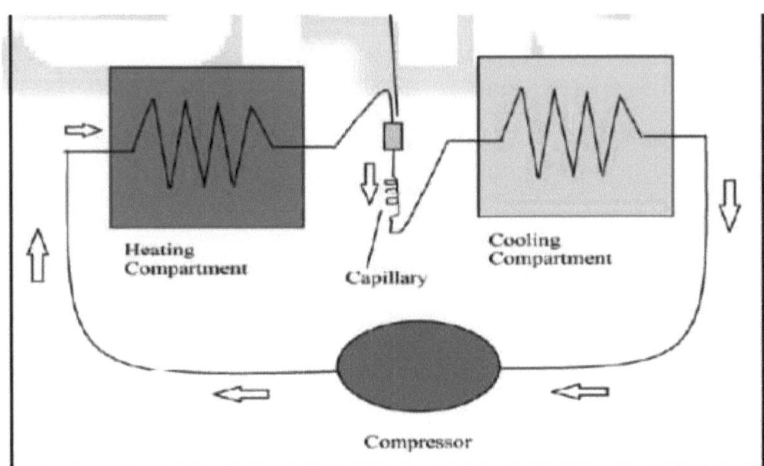

Figure 1. 16. Refrigerator operating cycle **[31]**.

They have shown that recovering waste heat from this refrigerator helps to reduce energy consumption, investment costs and gas emissions.

Pathak et al [32] studied the technique of waste heat recovery using a refrigeration system. They found that the refrigerator condenser provides a quantity of heat that influences the temperature in a room, especially during the summer. For this reason, they have used this waste heat to heat water and preheat food. The main advantage of this system is maximum heat production with minimum heat loss. They have shown that the performance of this system with the recovery of heat from the condenser is better than that of the conventional refrigerator. Thus, with the recovery of waste heat from the refrigerator condenser, the COP and efficiency are improved. They also concluded that this system can reduce energy and protect the environment.

Kumbhar et al [33] developed a technique for recovering waste heat from a domestic refrigerator. They showed that the recovery of waste heat gives several advantages: the performance of the system is improved, quantities of high-temperature water are produced, efficiency is increased, energy consumption and the cost of fuel for heating are reduced, and the quantity of water produced can be used for various applications.

Biswas et al [34] have recovered waste heat from domestic refrigerators to meet heating requirements. The aim of their work is to minimise the loss and maximise the amount of heat recovered by using water as a fluid for cooling the refrigerator condenser. They have confirmed that the recovered heat will be used in domestic applications such as cleaning, washing and drying. This heat recovery system can help to protect energy and reduce investment costs, as well as improving refrigerator efficiency.

Sreejith [35] experimentally studied a domestic refrigerator which uses water to cool its condenser with the use of different compressor lubricants. He found that the types of compressor lubricants used influenced the performance of the refrigerator under all loading conditions. They have also helped to reduce energy consumption while taking into account the influence on the COP of this domestic machine. He also found that recovering this heat from the condenser can produce a quantity of water equal to 200 litres at a temperature of

58°C per day. This heating technique was the main contributor to energy consumption.
Sreejith et al [36] experimentally studied a domestic refrigerator using the evaporator to cool the condenser. They compared the principle of cooling the refrigerator condenser with the use of the evaporator on one side and ambient air on the other. They showed that the performance of this system is improved by using the evaporator as the source of cooling, compared with using air. Similarly, using the evaporator to cool the condenser was shown to be the main factor in reducing energy consumption. The COP of this domestic machine improved when the evaporator was used to carry out the Freon condensation technique compared with the conventional one: condensation by ambient air. As a result, the efficiency of this domestic machine and the refrigeration effect are increased.
Tanmay Patil et al [37] studied the recovery of waste heat from the condenser of a domestic refrigerator. The domestic refrigerator is used to keep food in good condition. With the preservation of these foods, it consumed a significant amount of electrical energy. For this reason, the researchers decided to conserve this energy by recovering the heat lost from the refrigerator condenser. The various heat losses recovered by the condenser have been used to heat closed chambers or to heat water. The condenser, cooled by the ambient air, must supply a quantity of heat to the atmosphere. The condenser, on the other hand, is cooled by water. The amount of heat lost is already transferred to the water to bring it up to a high temperature. The low temperature obtained in the condenser can be used, for example , as an additional source of heat for preheating.
Tarang Agarwal et al [38] improved the COP of a domestic refrigerator with a condenser cooled by ambient air and using R134a freon as the refrigerant. They used a closed cabin mounted above the refrigerator to recover the heat lost by the condenser in order to heat the water. The main objective of this work is to minimise the heat lost and maximise the performance of this system. This system can be used as a conventional refrigerator if the cabin door is left open. They have shown that this recovery technique contributes to an increase in COP of around 11%.

4 Water desalination

Freshwater is an abundant natural resource, covering three quarters of the earth's surface. However, only around 3% of all water sources are drinkable. Around 25% of the world's population does not have access to fresh water of good quality and/or in large quantities. Lack of water will be a factor limiting food production. The need fresh water is considered an international problem, according to the World Water Council. 17% of the world's population will be living with little freshwater by 2020**[39]**. A gradual increase in the amount of water consumed as a result of population growth, improved living conditions and, not least, industrial and agricultural needs **[40].** Desalination of seawater or brackish water is the only solution that meets the needs of the population **[41].** It has become a reliable method of supplying drinking water throughout the world. The technique is already being used successfully and has proved more cost-effective **[42].**

4.1 World water resources

As Table 1.2 shows, the world's water resources are seas, oceans, glaciers, rivers, groundwater and lakes. However, freshwater represents only 2.5% of total water, and of the 2.5% of freshwater, lakes, rivers and groundwater represent 14%, or the equivalent of 0.35% of total water. 86% of the remaining freshwater is frozen at the poles.
Table 1. 2. Water distribution **[43]**

Source of water	Quantity (%)
Freshwater lakes	0.009
River water s	0.0001
Groundwater (near the surface)	0.005
Groundwater (at depth)	0.61
Water in glaciers and ice caps	2.15
Salt water from lakes or inland seas	0.008
Water in the atmosphere	0.0001
Ocean water	97.2

4.1.1 Distillable waters

Sea salinity varies from one sea to another, averaging 35 $g.L^{-1}$, with strong regional variations **[44]** in some cases: 39 $g.L^{-1}$ in the Mediterranean, 42 $g.L^{-1}$ in the Persian Gulf and up to 270 $g.L^{-1}$ in the Dead Sea (Table 1.3).

Table 1.3: Degrees of salinity in water **[45].**

Seas	Salinity (g.L-1)
Baltic Sea	7.0
Caspian Sea	13.5
Black Sea	13.0
Adriatic Sea	25.0
Pacific Ocean	33.0
Indian Ocean	33. 8
Atlantic Ocean	36.0
Mediterranean Sea	39.4
Arabian Gulf	43.0

4.1.2 Brackish water [46]

These are non-drinking waters whose salinity is lower than that of seawater and which can be classified into three categories:

Slightly brackish water 1 $g.l^{-1}$ to 5 $g.l^{-1}$,

Moderately brackish water 5 $g.l^{-1}$ to 15 $g.l^{-1}$

Very brackish water 15 $g.l^{-1}$ to 35 $g.l^{-1}$

4.1.3 Natural waters

This is the water that comes from lakes, rivers and groundwater. They have different chemical compositions and are sometimes polluted and unfit for consumption. They account for almost 14% of fresh water.

4.1.4 Waste water

This is water discharged by domestic, industrial or agricultural communities.

4.1.5 Drinking water

According to WHO (World Health Organisation) health standards, all water distributed to a community must be drinkable. Water is considered potable if its total salinity is between 100 and 1000 parts per million (i.e. 0.1 and 1 $g.L^{-1}$).

4.2 Desalination processes

Water desalination technology is a process removes the salt from seawater. The purpose of desalination is to fresh water for drinking or irrigation. There are two main desalination processes: membrane desalination and thermal desalination. The first method uses a special filter (membrane) to produce drinking water, while the second technology heats and vaporises seawater to obtain saturated water vapour and condenses it (under cooling) to finally distilled drinking water.

4.3 Desalination processes

Desalination processes fall into two main categories: thermal processes, also known as phase change processes, and membrane processes, also known as single-phase processes. Thermal processes evaporating salt water, which is then condensed to produce drinking water. This requires a thermal energy source such as solar, fossil and nuclear power. Thermal processes include successive expansion distillation (SED), multiple effect distillation (ME), fractional solidification or freeze distillation and solar distillation. The most widespread membrane processes on a commercial scale reverse osmosis and electrodialysis.

4.3.1 Membrane processes

4.3.1.1 Reverse osmosis

Reverse osmosis (RO) is a water purification process that uses a partially permeable membrane to remove unwanted ions and molecules from drinking water. In reverse osmosis, the pressure applied is greater than the osmotic pressure. Reverse osmosis can remove several types of dissolved and suspended chemical species, as well as bacteria, from water. The solute is retained on the pressurised side of the membrane, while the pure solvent passes through to the other side.

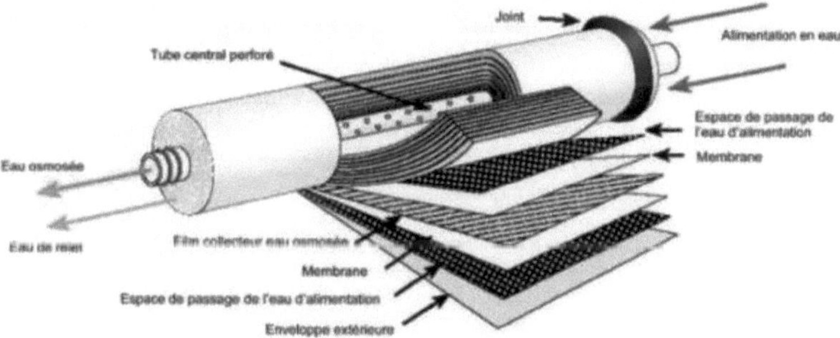

Figure 1. 17. Reverse osmosis module

4.3.1.2 Electrodialysis

Electrodialysis consists of transporting salt ions from one solution through ion exchange membranes to another solution under the influence an electrical potential difference. The electrodialysis cell consists of a dilute compartment and a concentrated compartment formed by an anion exchange membrane and a cation exchange membrane placed between two electrodes. In practice, several electrodialysis cells are grouped together to form an electrodialysis stack. The cations move in the direction of the electric current. They leave the first compartment through the cationic membrane, but cannot move on to the next compartment because the anionic membrane is an obstacle for them. The anions, on the other hand, move in the opposite direction to the electric current. They leave the first compartment by passing through the anionic membrane, but cannot pass to the second compartment because the cationic membrane inhibits their passage.

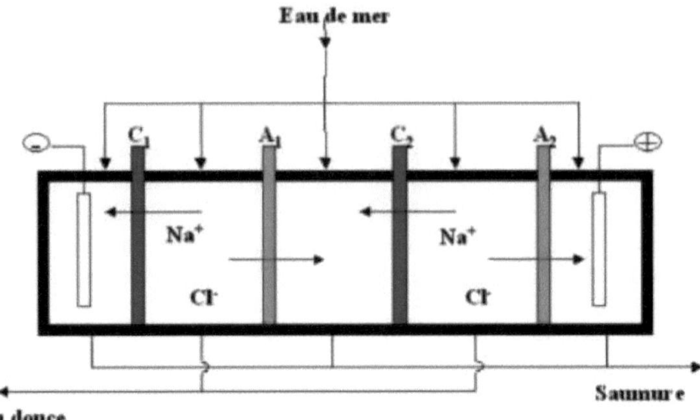

Figure 1. 18. Schematic diagram of the electrodialysis process

4.3.2 Thermal processes

4.3.2.1 Successive expansion distillation (FED)

This is a series of stages where each stage consists of a heat exchanger and a condensate collector. The multi-flash distiller has two ends, a hot end and a cold end. The stages have intermediate temperatures between the hot and cold ends. The stages have different pressures. Boiler water at high temperature is partially vaporised by expansion in a chamber. The steam produced is condensed in the condensation tubes and fresh water is collected.

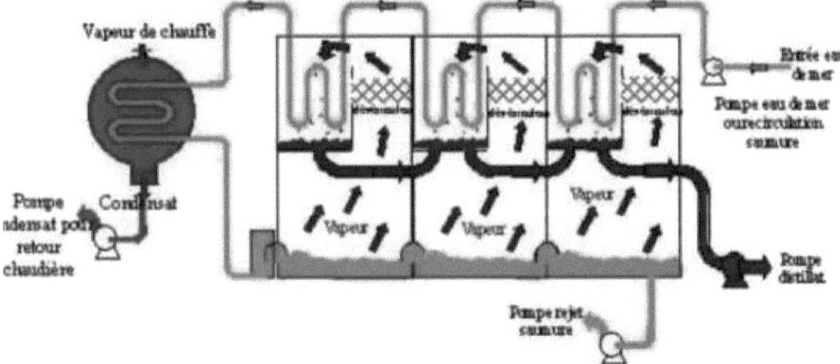

Figure 1. 19 MSF distillation

4.3.2.2 Multiple effect distillation (MED)

This system consists a series of evaporators, also known as effects. The vapour produced by the first evaporator is condensed in the second evaporator. The heat released during condensation enables the seawater in the second evaporator to evaporate, and so on for the other effects.

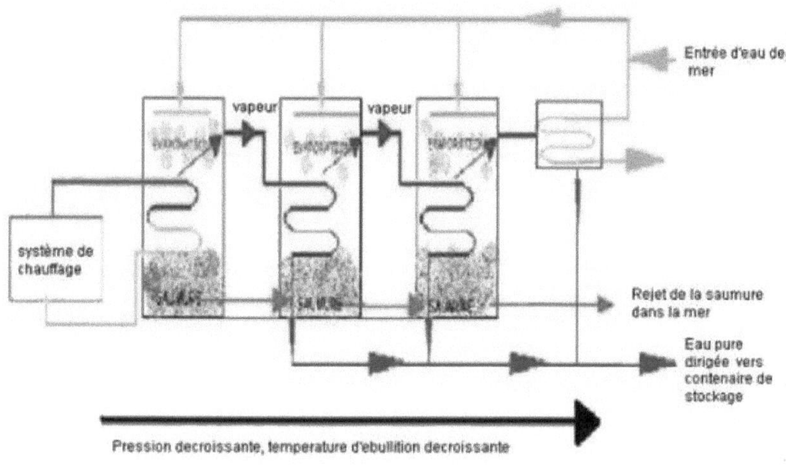

Figure 1. 20. MED distillation

4.3.2.3 Fractional solidification (freeze distillation)

Desalination of water by freezing also exists. This technique involves partially freezing salt water. The solidification temperature of pure water is higher than that of salt water. The ice crystals formed are then separated from the brine. Finally, the ice crystals are fused together to produce liquid drinking water.

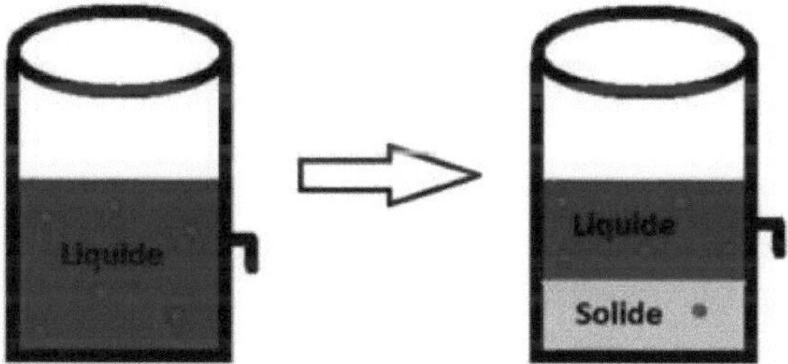

Figure 1. 21: Diagram of fractional solidification

4.3.3 Solar distillation

Solar distillation involves harnessing the sun's energy to evaporate water and collect its condensate. It differs from other desalination techniques that consume much more energy, such as reverse osmosis or distillation using fossil fuels, such as MSF distillation.

4.3.3.1 Single-acting solar distiller

The single-effect solar distiller is the original design of solar distillation. This type of simple solar distiller has a single pane of glass above the salt water, which wastes a lot of heat energy through the latent heat of condensation in the form of conduction through the glass. As a result, the efficiency of these solar distillers around 30 to 40% and a distillate production rate of around 6 l / m^2 / day **[47]**. In order to increase efficiency, several design modifications have been made. These various design modifications can be broadly classified into two groups: passive solar distillers and active solar distillers. Passive distillers use the internal heat of the distiller for the evaporation process, while active distillers use external energy sources, such as solar collectors or industrial waste heat. Passive single-effect solar distillers include tank-type solar distillers.

4.3.3.2 Passive solar distillers [47]

4.3.3.2.1 Solar tank distillers

This type of distiller consists of a black-painted basin in salt water emerges. The salt water is heated and then evaporated by the thermal energy of sunlight transmitted through a glass cover and then absorbed by the basin. The water vapour is then condensed on the glass and the distillate water is collected.

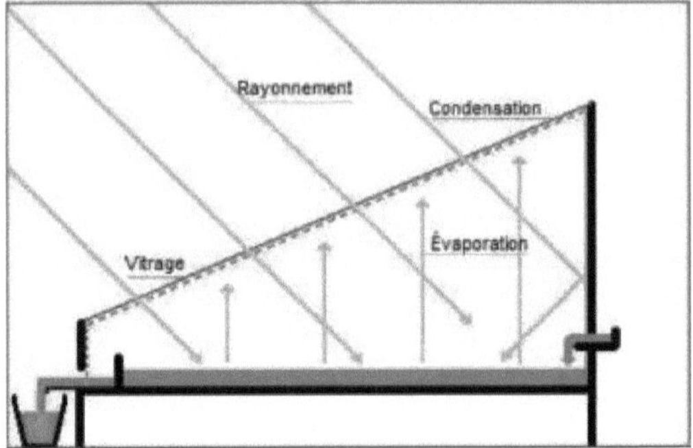

Figure 1. 22. Solar basin distiller

4.3.3.2.2 Solar diffusion distillers

In the simple diffusion desalination system, the hot and cold surfaces are placed parallel to each other and separated by a small distance. Air fills the space between the two surfaces. The gap is chosen to be small in order to eliminate convective heat transfer between the two surfaces. Solar radiation is transmitted through the glass and absorbed on the hot surface. When the feed water is allowed to flow onto the hot surface, water vapour diffuses across the gap where it is condensed on the cold surface. This process is called diffusion desalination.

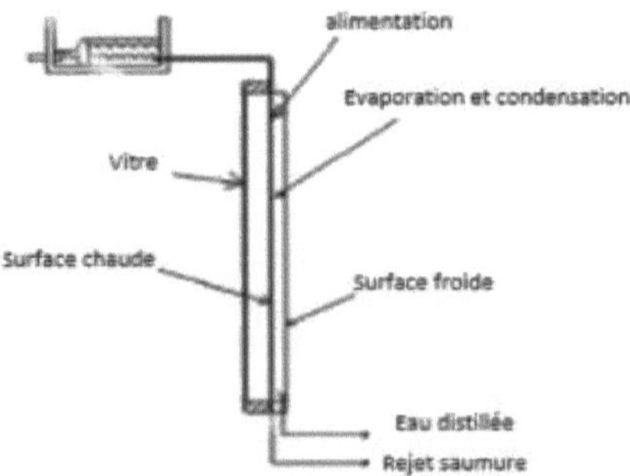

Figure 1. 23. Solar diffusion distiller

4.3.3.2.3 Spherical solar distillers

This type of distiller consists of a glass sphere and a black-painted metal plate placed horizontally in the centre of the sphere.

placed horizontally in the centre of the sphere. Water condenses on the surface of the glass lid of the glass lid and fresh water is collected in a tray at the bottom of the distiller.

This distiller is 30% more efficient than a conventional solar distiller.

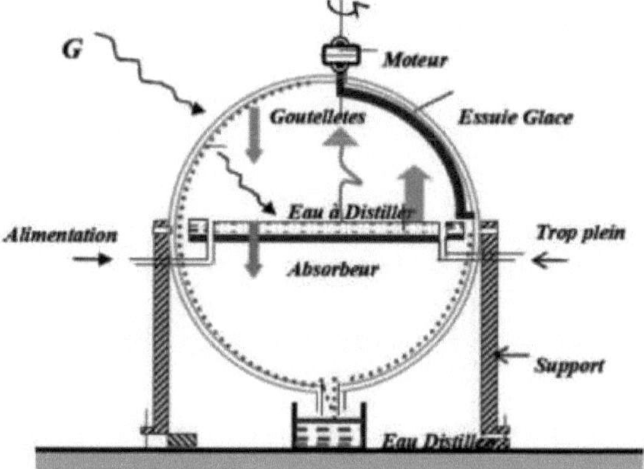

Figure 1. 24. Spherical solar distiller

4.3.3.3 Active solar distillers [47]

4.3.3.3.1 Integrated solar distiller with flat plate collector or solar concentrator

These types of solar distillers make it possible further increase the temperature of the salt water in the tank by means of a flat-plate solar collector or a solar concentrator, improving efficiency. An example of this type of distiller is shown in Figure 1.25 and Figure 1.26.

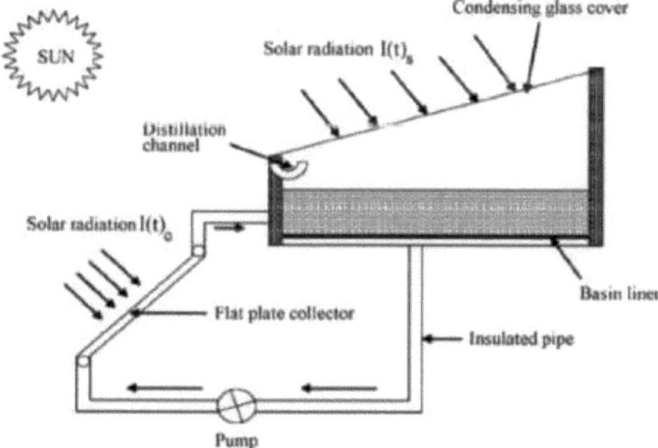

Figure 1. 25. Integrated solar distiller with plate collector

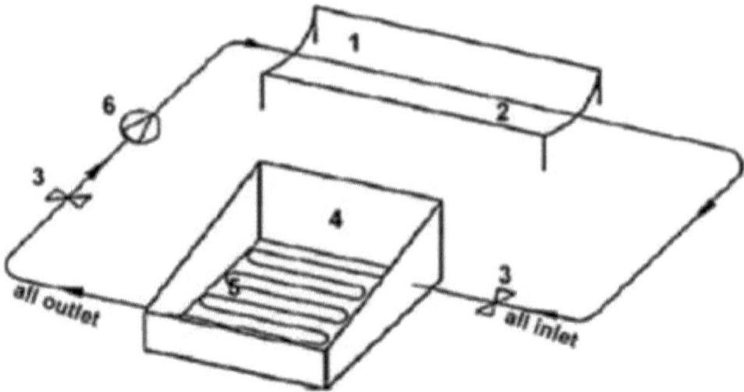

Figure 1. 26. Integrated solar distiller with solar concentrator

4.3.3.3.2 Active solar distiller with regeneration

The solar distiller with regeneration mainly consists of cooling the glazing, which heats up due to the latent heat of condensation, using a continuous flow of water above the glazing (Figure 1.27). The hot water leaving the cover is directly as a feed to the basin, raising the temperature of the water in the basin and thus increasing evaporation. This technique is called regeneration.

I and II Fiat-plate collector

III Tube-In-tube heat exchanger (Insulated)

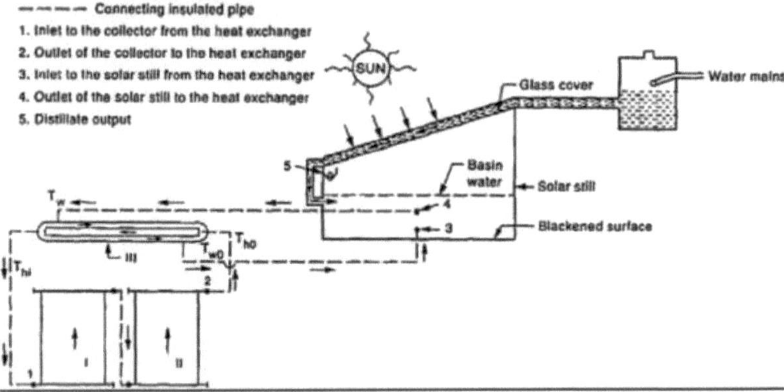

Figure 1. 27. Solar distiller with regeneration

4.3.3.3.3 Solar bubble distiller

This system uses a fan to create a bubbling effect in the distiller, which increases the rate evaporation. The bubbling effect created by the fan ensures that the raw water circulates and distributes the heat energy evenly. The evaporated raw water is condensed in the glazing in the form of drops. The air-blown solar distiller offers greater efficiency and higher output than a solar distiller without a fan.

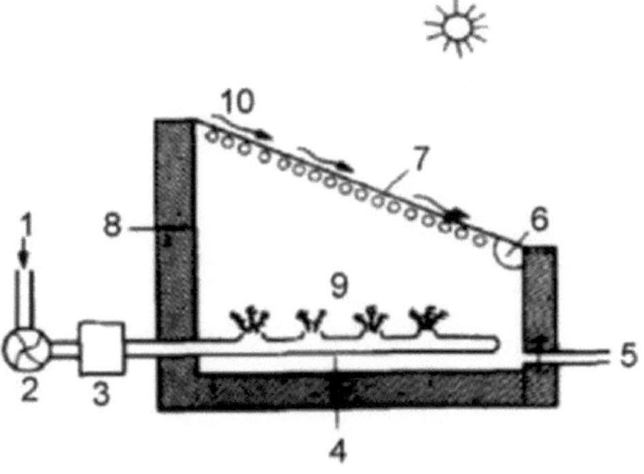

Figure 1. 28. Solar air bubble distiller

4.3.3.3.4 Solar distillation coupled with a hybrid photovoltaic and thermal system

Hybrid photovoltaic and thermal (PV/T) modules are photovoltaic modules coupled to heat extraction devices, which preserve the lifespan of photovoltaic cells overheated by solar radiation. The heat extracted from the modules can be used to help raise the temperature of the salt water emerging in the tank, thereby improving the efficiency of the distiller.

Figure 1. 29. Solar distiller coupled to a hybrid photovoltaic and thermal system

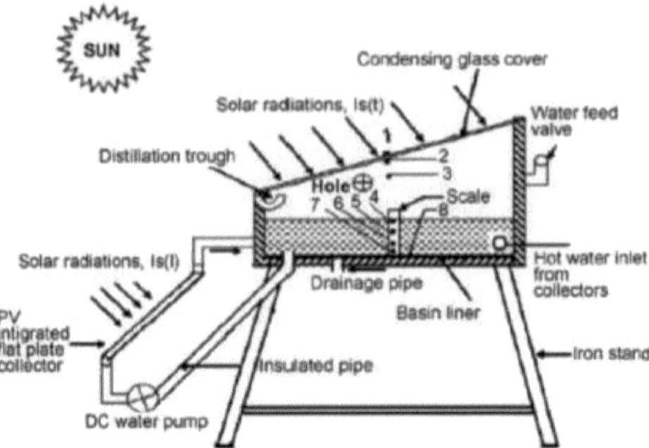

Figure 1. 30. Solar distiller coupled to a hybrid photovoltaic and thermal system

4.3.3.4 The multiple-effect solar distiller [47]

In triple basin solar distillers, two glass covers are fitted between the basin lining and the glass cover of the single basin distiller. These 2 glass covers serve as a second and third basin, also containing brine, placed one above the other. The total productivity of the system is then the sum of the productivities of the three basins.

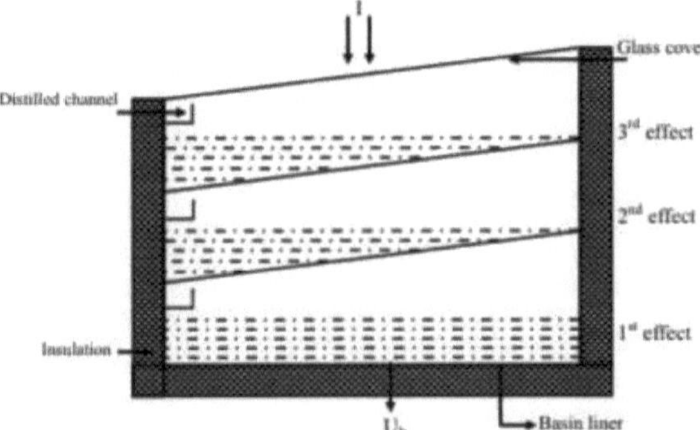

Figure 1. 31. Diagram of a triple tank solar distiller

4.3.3.5 Solar distiller coupled to a heat pump

This type of distiller is classified as a simple active solar distiller. The solar distiller coupled with a heat pump is part of this work. Adding a heat pump to a simple solar distiller improves efficiency. The system consists mainly of a tank containing brackish water, a glass cover and a compression heat pump. The compression heat pump consists of a condenser immersed in the water tank, an evaporator located under the upper part of the glass cover, a compressor and an expansion valve. The condenser will contribute to the formation of the water in the heat tank, and therefore to its evaporation, during the day and especially during periods of low irradiation by the refrigerant (R134a) passing through the heat pump. At , on the other hand, the evaporator will condense a large proportion of water vapour. The water in the pool is heated by incident solar radiation transmitted through the transparent glass cover and the condenser. Some of the water will evaporate and condense under the glass and evaporator. The condensate is then collected by two collectors.

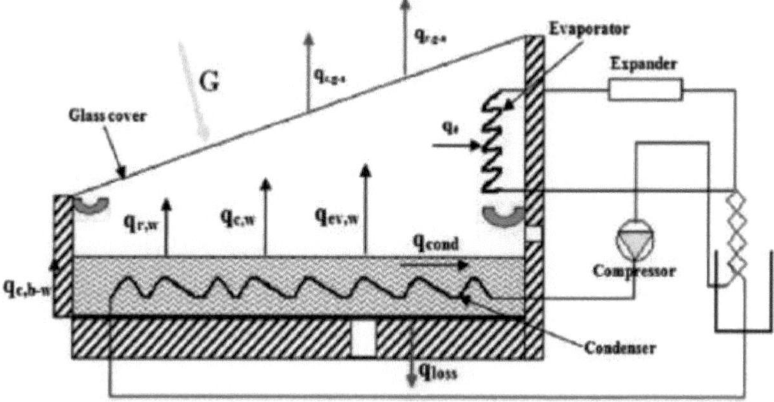

Figure 1. 32. Simple solar distiller coupled to a heat pump

A heat pump is a thermodynamic machine consisting of a closed circuit in which a refrigerant circulates. This circuit is made up of four main components: a compressor, an expansion valve, a condenser and an expansion valve. The role of this machine is to transfer energy from a cold medium to a warm medium. The refrigerant circulating in this circuit goes through a closed cycle made up of four stages. During these stages, the refrigerant will change state (liquid or vapour) and will be at different pressures and temperatures.

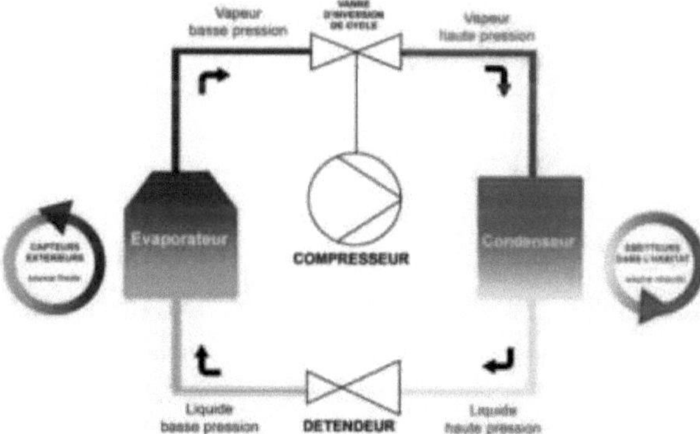

Figure 1. 33. Compression heat pump

The basic cycle of such a machine (with single-stage compression or a single compressor) can be broken down into four stages illustrated in an enthalpy diagram (Log P= g (H)) used by refrigeration engineers.

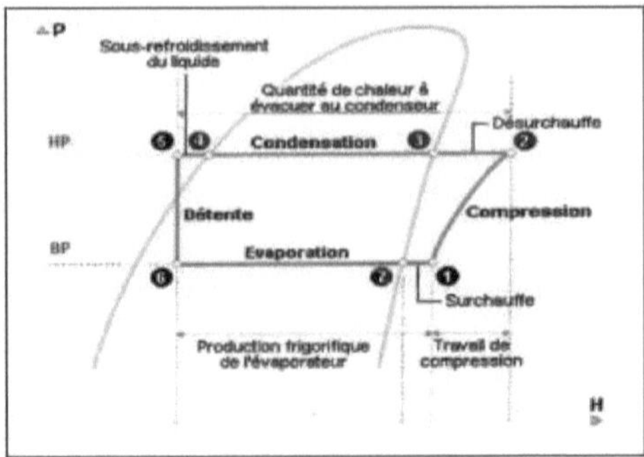

Figure 1. 34. Enthalpy diagram of a compression heat pump

4.4 Parameters influencing distiller operation [48]

The parameters that influence the operation of the distiller are classified into internal and external parameters.

4.4.1 Internal parameters

4.4.1.1 Glass

The main role of the pane is that it can be made of glass or plastic (Plexiglas, polycarbonate). It has two roles to play: firstly, it is a selective filter for solar radiation and secondly, it is a surface for condensation of water vapour. Good wettability is necessary to avoid condensation into droplets that tend to fall back into the pool and reflect a large proportion of the incident radiation, ensuring that the condensed water runs off towards the collector **[48]**.

4.4.1.2 Tilt

Its inclination to the horizontal determines the amount of solar energy introduced into the distiller. To minimise the distance between the brine and the glass, the angle of inclination must be chosen carefully. The angle of inclination also influences the energy balance equations for the distiller's various components. It depends on how the distiller operates during year **[48]**:

In summer operation, $\beta = \beta - 10°$.

In winter operation, we have $\beta = \beta + 20°$.

In annual operation, $\beta = \beta + 10°$.

4.4.1.3 Absorber

Studies carried out in this field show that the absorbent surface can be made of several materials (wood, metal, concrete, plastic or ordinary glass). The choice of material for the absorbent surface or black box depends on its thermal inertia, resistance to oxidation by water and mineral deposits.

4.4.1.4 Distance between evaporating surface and condensing surface

Hansen et al [49] found that distiller output increases as the distance between the brine and the glass decreases.

4.4.1.5 Thickness of the mass of water to be distilled

The thickness of the brine plays a very important role. Production is higher for a distiller with a low brine thickness, but for a distiller with a high brine thickness, maximum production is not observed until shortly after sunset.

4.4.1.6 Insulating the sides of the distiller

The purpose of insulating the sides is to eliminate heat loss to outside.

4.4.2 External parameters

The external parameters influencing the operation of the distiller are meteorological parameters:

4.4.2.1 Ambient air temperature

Research has shown that increasing the ambient temperature improves production.

4.4.2.2 Wind speed

Wind speed has the main effect on the convective exchange between the glass and the environment. This influences the temperature of the glass and therefore the efficiency.

4.4.2.3 Cloud intermittence and air humidity

Wind speed and ambient temperature depend on cloud intermittence, which explains the link with yield.

5 Conclusion

In this chapter, we have presented a general overview of compression refrigeration machines and heat pumps for the production of domestic hot water and the desalination of brackish water. The first part is devoted to describing the basic concepts needed to understand refrigeration machines and heat pumps. We have seen the advantages of this new technology for producing domestic hot water. Next, a description of the heat pump components was presented. The various parameters that influence the performance of the refrigeration machine, such as the geometry of the condenser, were also presented. In the second part, we presented the different desalination techniques available. In particular, we have shown that membrane processes, thermal processes such as desalination by successive expansion (MSF), multi-effect (MED) and vapour compression are expensive thermal desalination processes. We have also presented the parameters influencing the operation of the distiller. We have drawn particular attention to the possibility of using the heat given off by condensers in desalination processes. This bibliographical study shows the importance of using waste heat from refrigeration machines for water heating and desalination.

CHAPTER 2

Experimental study

1 Introduction

This chapter is devoted to the experimental study of two domestic refrigerators that heat water and an air conditioner that distils brackish water. The first part of this chapter is devoted to the domestic water heating part. The two prototypes used to produce domestic hot water are presented, along with the equipment and measurement methods used. The second part of the chapter presents the techniques for distilling brackish water using the heat released by the condenser of an air-conditioning unit.

2 Heating domestic water

2.1 Presentation of the prototype

A domestic refrigerator with a water heating unit is based on the same principle of the vapour compression cycle but with a small modification at the condenser, as shown in Figure 2.1. In this research framework, we only study the mode of domestic hot water production through the thermal discharge from the refrigerator condenser.

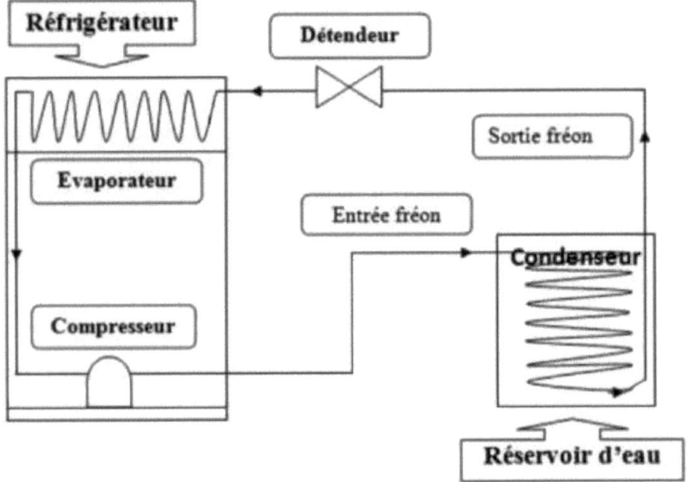

Figure 2. 1: Refrigerator used to heat water

2.2 Description of the experimental set-up

Two experimental devices have been built at the Gabès National Engineering School in the Energy, Water, Environment and Processes research laboratory. The first device is a mini fridge with a capacity of 46 litres coupled to a water tank. The second device is a domestic refrigerator with a capacity of 190 litres coupled to a cylindrical water tank containing a fully immersed helical condenser. Both refrigerators use R134a as the refrigerant. A detailed description of each appliance is given below.

2.2.1 Presentation of the first prototype

The experimental set-up consists of the standard components of a vapour compression machine. It comprises a condenser, a compressor, an expansion valve, an evaporator and a water tank, all of which are assembled together. A photo of this assembly is shown in Figure 2.2. The prototype used in this study is a Mini Fridge with a total net volume of 46 litres and a reversible door with a static cooling system. Like any conventional refrigerator, the condenser is air-cooled. In this new context, we have chosen to modify the condenser of the refrigerator cooled by the ambient air with another completely immersed in water, the aim of which is to make the most of the heat lost to the atmosphere. The main characteristics of the refrigerator used in the experimental study are given in Table 2.1.

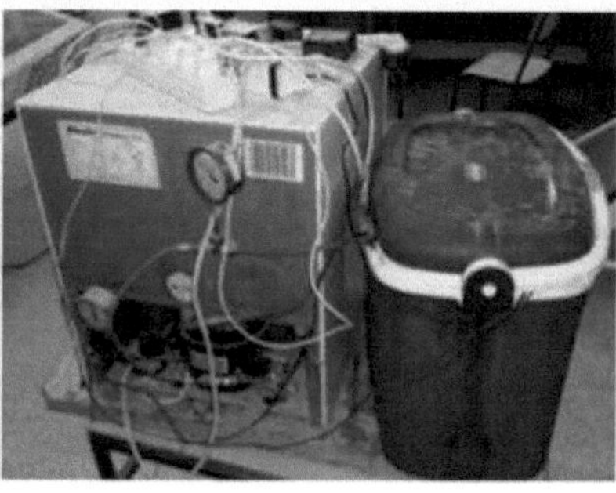

(a) Front view (b) Rear view

Figure 2. 2: Prototype of a mini fridge coupled to a water heater

Table 2.1: Specifications and features of the mini fridge with water heater

Fridge model	Compressor model : AES30DS
Climatic class: ST	Engine type: RSIR
Rated current: 1.2 A	Cooling capacity: 88 W
Refrigerant gas : R134a	COP: 2.98
Volume: 46 litres	Displacement: 3.88 cm^3
Energy consumption: 0.52 kWh/24h	Power supply: 220-240 V/ 50 Hz

2.2.2 Presentation of the second prototype

Figure 2.3 shows the prototype of a refrigerator that water. This assembly consists mainly of a tank in the form of a cylinder with a helical condenser immersed in the water, a compressor, an expansion valve and an evaporator. The tank in question contains 50 litres of water and has a cylindrical shape with a height of 480 mm and a diameter of 380 mm. The water tank is fully insulated with 50 mm thick glass wool wrapped around the plastic inner tank. The water tank is fully insulated to reduce heat loss. This water storage tank is fully protected by another steel tank as shown in Figure 2.3.

Figure 2.3: Domestic refrigerator with a fully immersed helicoïdal condenser Table 2.2 gives the specifications of the refrigerator coupled to a water heater used in this work.

Table 2.2: Characteristics of the refrigeration system

Components	Remarks
Fridge	Capacity: 190 L
Compressor	R-134a, Hermetically sealed, refrigerant gas: 140 g
Condenser	Length:13 m, copper
Evaporator	Type: natural convection
Water tank	Capacity: 50 litres, 50 mm of glass wool insulation

2.2.3 Presentation of the helicoidal condenser used in this study

To present the size and general shape of a water tank with a fully submerged helicoïdal condenser, we used a SolidWorks model design as shown in Figure 2.4.

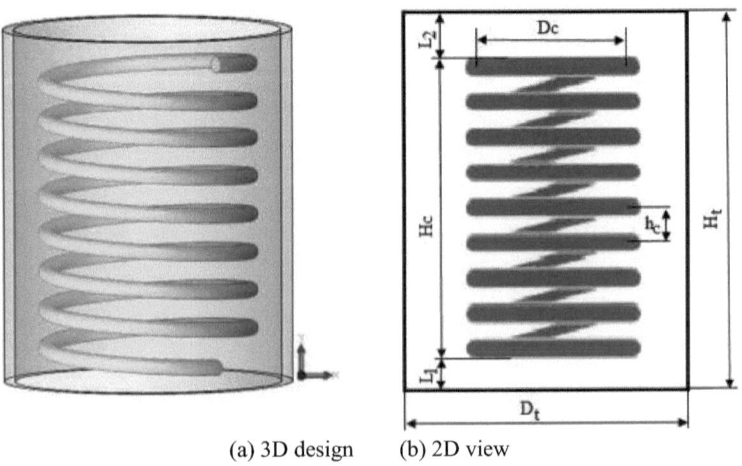

(a) 3D design (b) 2D view

Figure 2.4: The general shape of a helical condenser used in this work [50] The geometric parameters of the helical condenser tank are given in Table 2.3

Table 2.3: Geometric parameters of the water tank with helical condenser [50].

Component	Description	Symbol	Value
	Inside pipe diameter (m) Outside pipe diameter (m)	$d_{c,i}$	0.004
Condenser	Number of coils (-) Condenser height (m)	$d_{c,o}$	0.006
	Total length (m)	N	13
	Distance between turns (m)	H_c	0.38
		L_c	13.074
	Tank diameter (m) Tank height (m)	h_c	0.03
Tank water	Volume (litre)	D_t	0.38
		H_t	0.48
		V_t	50

2.3 Materials and methods

In terms of measuring equipment and methods, we chose two types of thermometer to measure different temperatures and two manometers to measure the pressure at the compressor inlet and outlet.

2.3.1 Instrumentation

To study experimentally the operation of the domestic refrigerator coupled to the water heater, several measurements of pressure and temperature were made on the various components of this system, as shown in Figure 2.5.

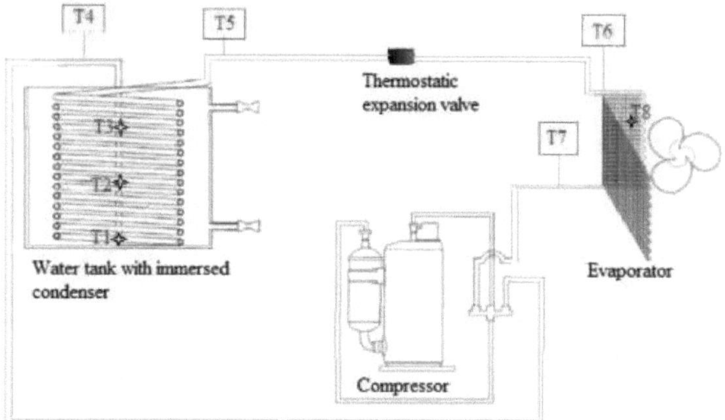

Figure 2. 5: Water cooling and heating system [50].

The measurements taken during the experiments are :

- The temperatures of the various refrigerator components, water and ambient temperature using thermocouples and thermometers.
- The pressures at the inlet and outlet of the refrigerator compressor for producing hot water using the manometers

2.3.1.1 Thermocouples

The temperatures at the various points are determined using a pre-calibrated type-K thermocouple. A minimum of seven thermocouples are used to determine the temperatures of the refrigerator components and the water at each point, as shown in Figure 2.5.

2.3.1.2 Thermometers

Two thermometers are used to measure the temperature inside the fridge and the ambient temperature.

2.3.1.3 Pressure gauges

The high and low pressures at the compressor inlet and outlet are measured using the pressure gauge.

2.3.1.4 Power consumption

A wattmeter is used to measure refrigerator's power supply.

2.3.2 Calibration of thermocouples

As shown in Figure 2.6, the calibration of the various thermocouples was carried out in comparison with the standard platinum resistance thermometer (SPRT) connected to measurement electronics with an absolute accuracy of ± 0.1°C. The electromotive forces generated by the thermocouples are conditioned by deviated analogue SB modules with an accuracy of ±0.05%. The temperature range explored varies from -50 °C to 150 °C and the temperature measurement accuracy is of the order of ±0.67 °C .

Figure 2.6: Calibration of thermocouples in ambient air

2.3.3 Location of thermocouples in the tank

For each test, three thermocouples were integrated in the vertical direction of the tank in order study the variation in water temperature inside the tank, as shown in Figure 2.7. Details of the locations of the thermocouples are shown below.

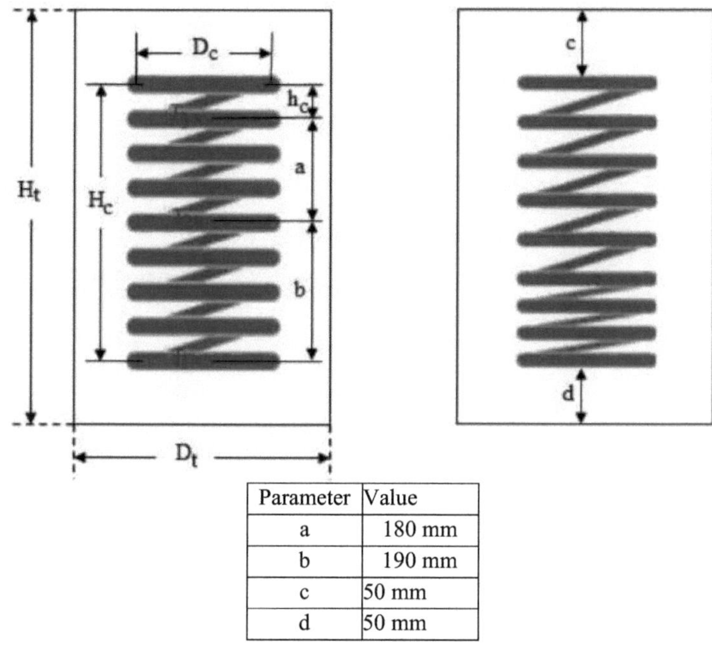

Parameter	Value
a	180 mm
b	190 mm
c	50 mm
d	50 mm

Figure 2. 7: Location of thermocouples inside the tank

2.3.4 Refrigeration plant measurement points

A minimum of four type K thermocouples are used to measure the refrigerant temperatures at the inlet and outlet of the condenser and evaporator as shown in Figure 2.8. The thermocouples are placed on the outer surface of the tube of each heat exchanger and allow the variation in temperature to be measured at any time.

The various temperatures and pressures measured in the refrigeration plant are as follows:

T_4 : R134a refrigerant temperature at condenser inlet (°C)

T_5 : R134a refrigerant temperature at condenser outlet (°C)

T_6: R134a refrigerant temperature at evaporator inlet (°C)

T_7: R134a refrigerant temperature at evaporator outlet (°C)

T_8: Air temperature inside the refrigerator (°C)

T*,^ 2*, ^3*: Water temperature in the different levels of the tank (°C)

P_1: R134a refrigerant pressure at compressor inlet (bar)

P_2: R134a refrigerant pressure at compressor outlet (bar)

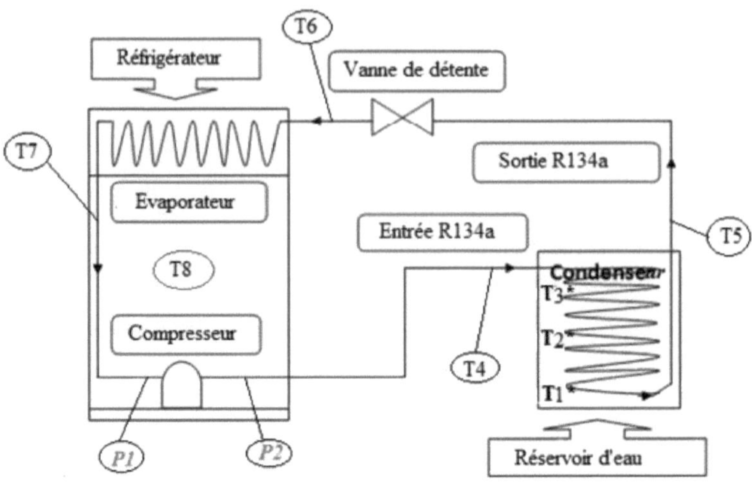

Figure 2.8: Temperatures and pressures measured in the refrigeration plant

The pressure given by a pressure gauge is a relative pressure. The absolute pressure is calculated using the following equation:

$$P_{abs} = P_{rel} + P_{atm} \quad (2\text{-}1)$$

Under these conditions, it is assumed that $P_{atm} = 1\ bar$

2.3.5 Error

During experimentation, errors in results and uncertainties may result from instrument calibration, operating conditions, choice of instrument and reading errors. Performance measurement errors can be calculated using the error propagation method reported by **Kline and McClintock [51]**. It can be presented as follows:

$$W_R = \left[\left(\frac{\partial R}{\partial x_1} W_1\right)^2 + \left(\frac{\partial R}{\partial x_2} W_2\right)^2 + \cdots + \left(\frac{\partial R}{\partial x_n} W_n\right)^2\right]^{1/2} \quad (2\text{-}2)$$

Where W_R is the uncertainty of the result, w_i, W2,..., w_n are the uncertainties of the independent variables. R is a given function of independent variables xi,x2...xn Using the equation proposed by Kline and McClintock, the maximum performance errors are i.6 %.

In order to avoid dispersion in the results, a series experiments was carried out on the same operational condition to show the possible variation in the experimental results if the same parameter is given. It should be noted that each test is repeated three times for the same variable to confirm the accuracy of the results. The uncertainty of the measurements is shown in Table 2.4.

Table 2.4: Measurement uncertainties

Parameters	Uncertainty	Measurement margin
Temperature (thermocouple)	±0.67	-50.0 to 150 °C
Temperature (Mercury thermometer)	-	-10 to 100 °C
Pressure gauge (for low pressure)	±0.1	0-2.0 MPa
Pressure gauge (for high pressure)	±0.1	0-3.0MPa
Performance	1.6 %	-

3 Distillation of brackish water

3.1 Description of the test bench

An air conditioner coupled to a water desalination unit is based on the same principle of the mechanical vapour compression cycle. The modification is in the external geometry of the condenser. The condenser cooled by the ambient air is replaced by another immersed in water to recover the heat rejected into the atmosphere.

The main objective of this study is to make the most of the heat supplied by the condenser to distil the water, as shown in Figure 2.9.

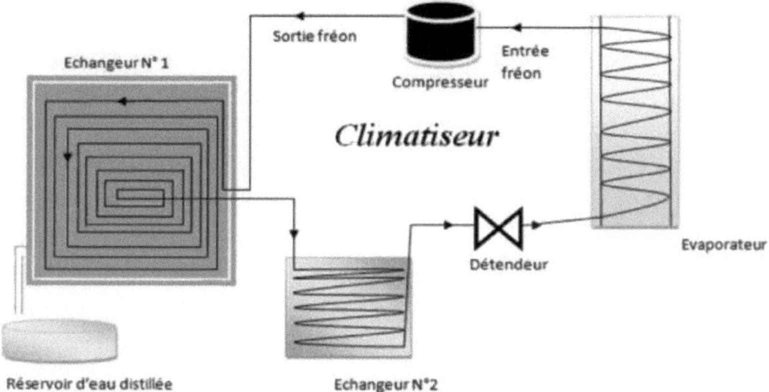

Figure 2. 9: Installation principle

The main components of the system are a tank containing brackish water, a glass cover and a compression heat pump. The compression heat pump consists of a condenser completely immersed in the water tank, an evaporator located in front of the water tank, a compressor and an expansion valve as shown in Figure 2.10. The condenser contributes to the formation of heat in the water of the basin, and therefore to its evaporation, during the operation of the system by the refrigerant (R134a) passing through the heat pump. On the other hand, the transparent glass cover will condense a large proportion of water vapour. Some of the water evaporates and condenses under the glass and the condenser. The distilled water is then collected in a collector.

Figure 2. 10. Air conditioner coupled to a distillation system

3.2 Measuring equipment

Measures taken during our experiment concerning :

- The temperatures of the various parts of the distiller, i.e. the glass, the basin, the water and the ambient environment. These values are taken using thermocouples and thermometers.
- The volume of distillate collected using a test tube.
- The pressures at the inlet and outlet of the heat pump compressor are measured using pressure gauges.

The tests are carried out in October and December 2019. Measurements are taken every hour throughout the day.

4 Conclusion

In this chapter, we have presented the prototypes of water-heated refrigerators and an air-conditioning unit for water desalination, as well as the measurement methods and equipment used in our study. We also focused on the calculation of uncertainties and measurement errors. In this way, we emphasise the fact that the amount of heat rejected into the atmosphere can be recovered and used in various other applications. The next chapter will be devoted to modelling the operation and coupling of the refrigerator to the water heater.

CHAPTER 3

Modelling the operation and coupling of the refrigerator to the water heater

1 Introduction

In this chapter, a model of the operation and coupling of the domestic refrigerator to the water heater is presented. The basic techniques for developing a simulation with ANSYS Fluent are described. We also present a mathematical model for the various components of the refrigerating machine coupled with a water heater. Next, we develop a theoretical model capable of determining the quantity of water to be heated daily by the thermal discharge from the condenser. Finally, we present the numerical results of the 2D model that we have chosen to model the water tank with a fully immersed helical condenser, comparing it with the experimental results that we have been able to develop in our Energy, Water, Environment and Processes research laboratory at the National Engineering School in Gabès.

2 Digital model

In this section, we present an overview of the ANSYS Fluent software that we used to develop our numerical simulations.

The numerical calculation was carried out using the ANSYS Fluent version R16.2 calculation code, which uses the finite volume method. We will briefly introduce the methodology for numerically solving the model with the "FLUENT" code, as shown in Figure 3.1.

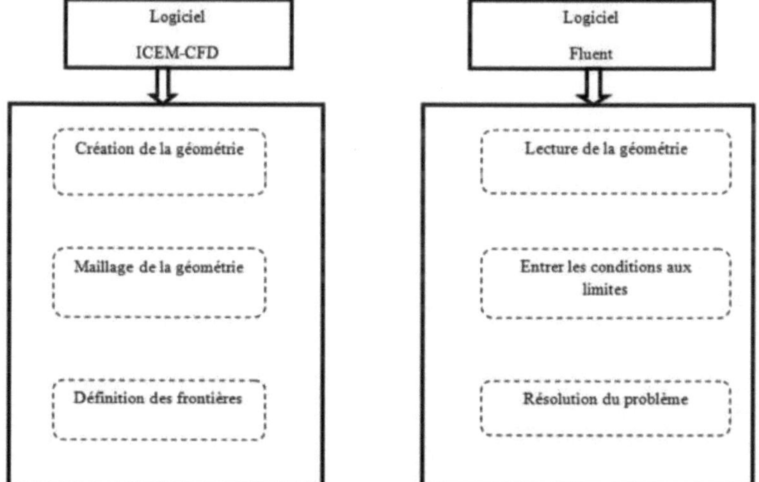

Figure 3.1: Overall numerical procedure for the simulation using the ICEM-CFD and Fluent calculation codes

2.1 Mesh generation

In CFD simulation, mesh generation is a primitive and complex operation. Most of the time, it is complicated to generate a mesh that is compatible with the exact geometry. Therefore, modification of the complete shape of the geometry to a certain level of acceptance is allowed until it does not affect the essential characteristics during CFD analysis. For numerical simulation, the creation of the geometry and the meshing are carried out by the ICEM-CFD.

The ICEM-CFD provides a second alternative for meshing, which is necessary to produce a good quality mesh for a complex geometry. ICEM-CFD is a subset of the commercial ANSYS-Fluent software. It is a complete meshing tool capable of producing complex geometry with intricate detail. It allows meshing for a given geometry with tetrahedral and hexahedral meshes, with the possibility of exporting a mesh file for varieties of CFD software.

2.2 CFD code

ANSYS FLUENT is an adaptive Computational Fluid Dynamics (CFD) modeller that supports heat transfer simulation and fluid flow visualisation.

This calculation code has high-performance resolution capabilities and can numerically study three- and two-dimensional geometries, turbulent, transient and laminar flows, and compressible and incompressible fluids. Fluent is also capable of producing flows in the liquid or gaseous state while being able to modify the properties of the fluid/solid. The Fluent calculation code uses the finite volume method. It deals with the equations governing the conservation of mass, momentum and energy. The discretisation phases are as follows:

> Division of the geometry into discrete control volumes using a computational grid. Integration of governing equations over individual control volumes to develop algebraic equations for unknown dependent variables, such as temperature, velocity and pressure ...

> Linearisation of discretised equations and solution of the resulting linear equation system.

2.3 Numerical resolution procedure

Once the two-dimensional geometry has been created and the boundaries indicated, the mesh is exported so that it can be digitally resolved and the integral equations defining the conservation of mass, momentum and energy can be discretised.

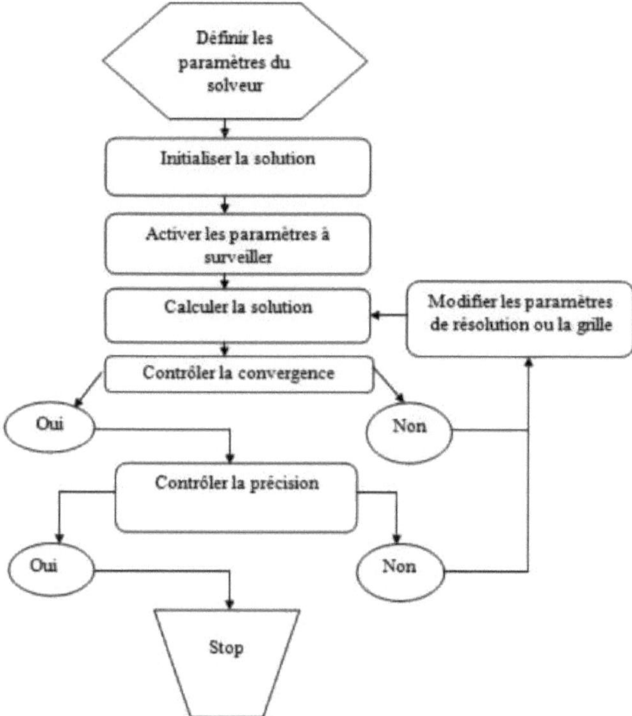

Figure 3. 2: Numerical resolution stage using ANSYS Fluent

2.4 Choice of solver formulation

ANSYS Fluent includes two types of solvers: the pressure-based solver and the density-based solver. The first solver is originally used for incompressible and weakly compressible fluid flows. The second is chosen for high-speed compressible fluid flows. Although both solvers are well developed for guiding a wide range of flows, the density-based solver may still have an advantage over the pressure-based solver for high-speed compressible flows.

2.4.1 Pressure-based solver

In ANSYS Fluent, there are two types of pressure-based numerical solution scheme: The Pressure-Based Segregated Algorithm and The Pressure-Based Coupled Algorithm. The first deals with the conservation of mass equation, the momentum equation and the energy equation, which are separated from each other, and the other solves the governing equations coupled to each other. The separate algorithm is more efficient because it can store the discretised equations once in memory. Convergence of the results is relatively slow because the equations are determined in a decoupled manner. The coupled algorithm memory space, but convergence is much better than with the coupled solver **[52]**.

2.4.2 The density-based solver

The density-based solver treats the governing equations in a coupled manner. The coupled system of equations can be treated using either an explicit coupled formulation or an implicit coupled formulation. The implicit formulation uses the existing and unknown values in neighbouring cells to determine the unknown value in a specific cell for a given variable. The equations must then be processed simultaneously as the various values appear in multiple equations of the system.

The explicit formulation uses only the current value, which means that the equations do not have to be set simultaneously.

2.5 Discretisation scheme

The discretisation of the equations involves transforming these differential equations into form of algebraic equations using approximations of the derivatives. The field variables stored in the cell centres must be interpolated onto the faces of the control volumes as shown in the following equation **[52]** :

-ᵛ + ΣN faces pfVf0f.Af= ΣN faces r0V0f. Af + S_0V (3-1)

Where ρ is the density, 0 is the time-averaged variable, *s0* is the source term and t is time.

Fluent uses several interpolation schemes for discretisation, namely

- *First-Order* Upwind diagram
- *Second-Order* Upwind diagram
- *Power Law* diagram
- QUICK "*Quadratic Upwind Interpolation for Convective Kinetics*" scheme
- *Bounded Central* Differencing scheme
- *Monotone Upstream-centered Schemes for Conservation Laws*".

2.5.1 Face pressure interpolation methods

The pressure interpolation at the cell faces provided by ANSYS Fluent are: ❖ The "*Standard*" scheme. ❖ The "*PRESTO*" scheme.

- The Linear diagram.
- The *Second-order* diagram.
- The *Body Force Weighted* diagram.

2.5.2 Choice of interpolation methods (Gradients)

Variable gradients are essential for estimating diffusive flows, velocity drifts and for high order discretisation parameters. The gradients of the variables on the cells are determined using a multidimensional Taylor series. Gradients are evaluated in ANSYS Fluent using the following methods **[52]:** ❖ Green-Gauss cell-Based

- Green-Gauss Node-Based
- least-squares cell-Based

2.5.3 Choice of pressure-velocity coupling method

Solving the equation for conservation of momentum and conservation of mass requires a coherent pressure field and velocity field at all times.

Four schemes are available on ANSYS Fluent **[52]**:

- *SIMPLE "SEMI-IMPLICIT Method for PRESSURE-LINKED Equations":* Default algorithm.
- *SIMPLEC "SIMPLE-CONSISTENT":* ensures faster convergence for simple models (laminar flows without the use of physical models).
- *PISO "PRESSURE-IMPLICIT with Splitting of Operators":* Useful for meshes containing nodes with a higher than average "highly skewed" asymmetry or for unsteady flows.
- *FSM "Fractional Step Method"*: is proposed for unsteady flows. It is used with the NITA algorithm and gives characteristics identical to those of the PISO scheme.

2.5.4 Initialization

With the right choice of boundary conditions, convergence is accelerated and the solution is stable.

2.5.5 Convergence criterion

This criterion is a specific case for residuals that specify the convergence of an iterative result. Convergence has been assessed on the basis of three rules. Firstly, the residuals for the momentum, continuity and volume fraction equations. They are monitored and should preferably fall below $10^{(-6)}$ **[50]**. Under some conditions, the residual criterion may never be met even if the solution is valid, and for others, the solution may be incorrect even if the residuals are low.

Equation (3-2) describes the calculation residual R **[52]** :

R = ^|ªw w$+ ªE E$+ Su - ªp0p| (3-2)

2.6 Mathematical model

In order to study the performance of a mechanical vapour compression refrigeration machine coupled with a hot water production system, we not only experimental studies but also numerical simulation. The latter offers the flexibility needed study the system's performance and the effect of the various geometric parameters of the condenser and tank. It is also possible to monitor the distribution of water temperature and velocity in the tank during the heating period. To achieve these objectives, several techniques can be considered for modelling a refrigeration machine coupled with a water heating unit.

2.6.1 Domestic refrigerator operating model

Studies concerning a domestic refrigerator coupled with a water heater are often accompanied by several models. Some simulate the operating principle of the complete system and others simulate each component independently of the rest of the equipment. In this section, we look at the modelling of the evaporator, condenser, expansion valve and compressor in the mechanical vapour compression cycle, as well as the modelling of the water storage tank.

2.6.1.1 Evaporator model

The evaporator is the main component of a refrigeration system. It is placed in the medium to be cooled. The medium emits heat at a higher temperature than the refrigerant. Thermal energy is therefore transferred from the medium to be cooled to the refrigerant. The refrigerant inside the tube heats up, causing it to vaporise.

- Evaporator operating condition

At the inlet to the heat exchanger (evaporator), the R134a refrigerant is in the state of a liquid-vapour mixture (two-phase flow) at low pressure and low temperature. At the outlet, the refrigerant is a superheated vapour (single-phase flow) at low pressure, as shown in Figure 3.3.

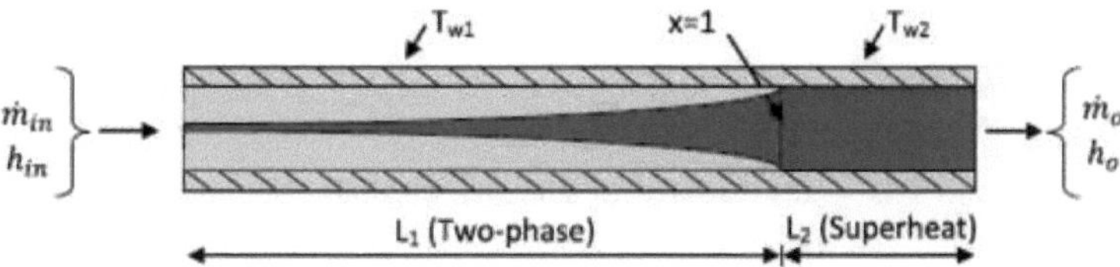

Figure 3.3: Evaporator with two refrigerant flow regimes **[2]**.

Consequently, the evaporator will be modelled according to these two types of flow (single-phase and two-phase).

For each type of flow, the power received by the refrigerant (R134a) at the evaporator is given by :

$$Q_e = \dot{m}_r \cdot (h_{e,r,o} - h_{e,r,i}) \quad (3\text{-}3)$$

The amount of energy lost by the external fluid (air) is gained by the refrigerant (R134a) and determined by the following equation:

$$Q_a = \dot{m}_a \cdot (h_{e,a,o} - h_{e,a,i}) \quad (3\text{-}4)$$

Generally speaking, the total amount of energy to be transferred is given by the following relationship

$$Q_e = Q_a = U_e \cdot A_e \cdot \Delta T_e \quad (3\text{-}5)$$

2.6.1.2 Compressor model

The main function of the compressor in the vapour compression refrigeration cycle is to draw in the refrigerant in the form of a superheated vapour at low pressure and low temperature and to discharge it at high pressure and high temperature to a heat exchanger (condenser). In this study, we will assume that the compressor is isentropic (reversible adiabatic).

When modelling the compressor, there are three important parameters to determine: refrigerant mass flow rate (m_r), electrical power consumption (W_{co}) and efficiency. (η_{co}).

The refrigerant mass flow rate at the compressor outlet is calculated as follows:

$$\dot{m}_r = \eta_v \cdot \frac{V_h}{3600 v_s} \quad (3\text{-}6)$$

Where $\eta_v$is the volumetric efficiency, ^is the compressor displacement and V_s is the suction mass volume.

The volumetric efficiency of the compressor is calculated by the following equation :

$$\eta_v = \frac{V_{c,r}}{V_{c,t}} \quad (3\text{-}7)$$

Where $V_{c,r}$ is the actual compression volume and $V_{c,t}$ is the theoretical compression volume.

The following equation is used to determine the electrical energy consumed by the compressor:

$$P_c = \frac{\dot{m}_r \cdot (h_{c,r,f} - h_{c,r,i})}{\eta_c} \quad (3\text{-}8)$$

Compressor efficiency is calculated using the following formula:

$$\eta_c = \frac{P_{ff}}{P_{ec}} \quad (3\text{-}9)$$

Where P_{ff} is the power supplied to the refrigerant and P_{ec} is the electrical energy consumed. Under these conditions, the power supplied to the refrigerant is calculated by the following formula:

$$P_{ff} = m_r \cdot (m_{c,r,f} - m_{c,r,i}) \quad (3\text{-}10)$$

2.6.1.3 Regulator model

The expansion valve is a device that isenthalqily expands the refrigerant leaving the condenser to a liquid state. On leaving the expansion valve, the refrigerant will be in the state of a liquid-vapour mixture (two-phase flow). In this study, we will consider that the expansion process is isenthalpic; the enthalpy at the inlet and outlet of the expansion valve are equal.

$$h_{ev,r,i} = h_{ev,r,o}$$

(3-11)

- **Principle of the model**

For the modelling, we chose a capillary-type expansion valve, which is a cylindrical device with a very small diameter and a long length. Consequently, with use a very long tube with a small diameter, there are two fundamental reasons that produce pressure losses in the tube:

- Since the diameter of the tube (pressure reducer) is very small and the length is very long, pressure loss occurs.
- The saturated liquid (single-phase flow) leaving the condenser is transformed into a liquid-vapour mixture (two-phase flow) when it enters the evaporator. Since the mass flow rate of the refrigerant is constant, its density decreases due to the formation of vapour. This leads to a very high velocity inside the tube and also a loss of pressure through acceleration occurs.

Figure 3.4 shows a capillary-type expansion valve connected between the condenser outlet and the evaporator inlet. The refrigerant flow inside the capillary tube divided into two flow regimes: single-phase flow (subcooled liquid phase) and two-phase flow (liquid-vapour phase).

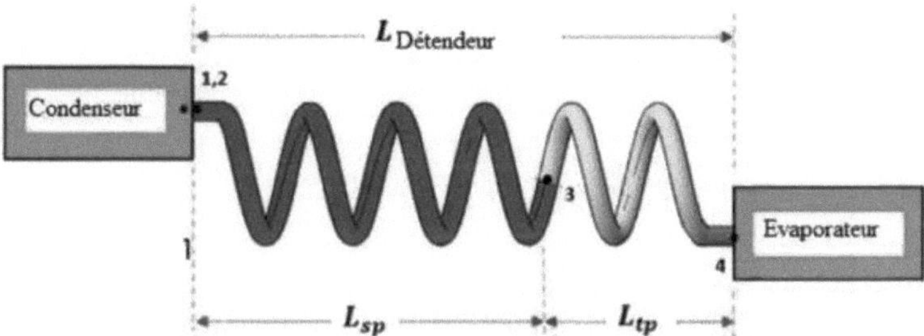

Figure 3.4 Refrigerant flow regime in the capillary-type expansion valve **[53].**

2.6.1.4 Condenser model

The condenser is a heat exchanger located in the external environment at a lower temperature than that of the refrigerant. It transfers heat energy from the refrigerant to the outside environment. The refrigerant will cool down, causing it to condense. Four temperature levels are obtained at the condenser: the inlet and outlet temperatures of the refrigerant (T_{ce} and T_{cs}) and the inlet and outlet temperatures of the cooling medium (T_{fe} and T_{fs}) as shown in Figure 3.5.

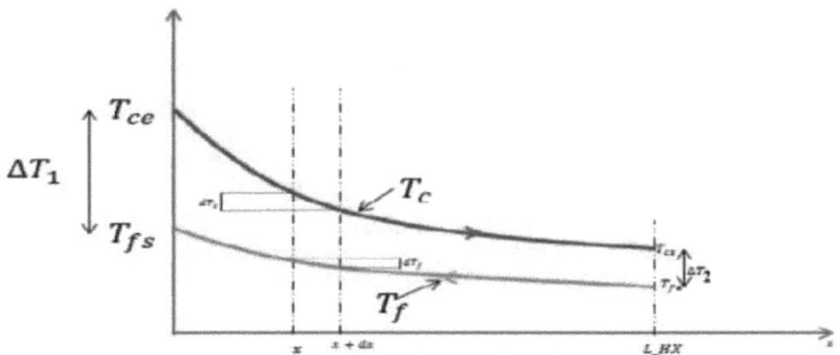

Figure 3.5: Diagram of the variation in refrigerant and external fluid temperatures along the

condenser.

Depending on the phase change of the refrigerant inside the condenser, the latter is divided into three different zones: superheated vapour, mixture of a liquid with vapour and subcooled liquid, as shown in Figure 3.6.

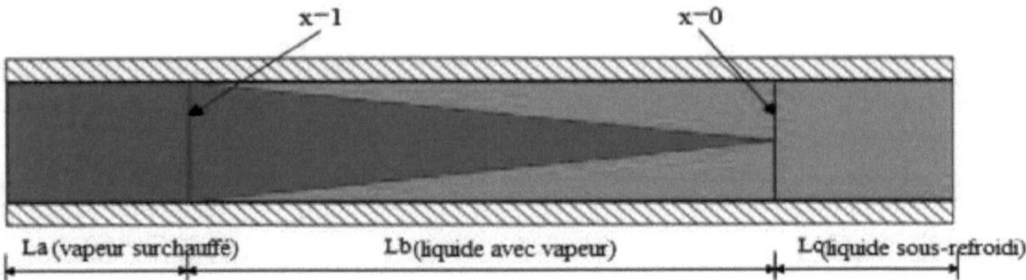

Figure 3.6 Diagram of a condenser with three refrigerant zones **[9].**

2.6.1.4.1 Modelling and calculation method

For each zone of the condenser, the energy conservation equation is described as follows:

$$Q_c = \dot{m}_r \cdot (h_{c,r,i} - h_{c,r,o}) \quad \text{(3-12)}$$

The energy transferred to the water is calculated using the following formula:

$$Q_w = Cp_w \cdot \dot{m}_w \cdot (T_{w,o} - T_{w,i}) \quad \text{(3-13)}$$

The heat transfer equation is as follows:

$$Q_w = Q_c = U_c \cdot A_c \cdot (T_{c,r} - T_w) \quad \text{(3-14)}$$

Where U_c is the total heat transfer coefficient, A_c is the surface area of the condenser, $T_{c,r}$ is the temperature of the refrigerant in the condenser and T_w is the temperature of the water. The total heat transfer coefficient is determined from the following equation :

$$U_c = \left(\frac{d_{c,o}}{\alpha_{c,r} d_{c,i}} + \frac{1}{\alpha_w} + \frac{d_{c,o}}{2\lambda_c} \ln\frac{d_{c,o}}{d_{c,i}}\right)^{-1} \quad \text{(3-15)}$$

Where a_w is the heat transfer coefficient on the water side, $a_{c,r}$ is the heat transfer coefficient on the refrigerant side, $d_{c,i}$ and $d_{c,o}$ are the inside and outside diameters of the tube and λ_c is the thermal conductivity of the copper tube.

The water-side heat transfer coefficient can be calculated from this theory:

$$\alpha_w = \frac{Nu_{w,d} \cdot \lambda_w}{d_{c,o}} \quad \text{(3-16)}$$

Where $Nu_{w,d}$ is the Nusselt number. The Nusselt number can be calculated based the outside diameter of the condenser tube and is given by :

$$Nu_{w,d} = 0.290.(Ra_{w,d})^{0.293} \text{ pour } 4 \times 10^3 \leq Ra_{w,d} \leq 1 \times 10^5 \quad \text{(3-17)}$$

The Rayleigh number can be written as :

$$Ra_{w,d} = \frac{g\beta\Delta T_w d_{c,o}^3}{v_w a_w} \quad \text{(3-18)}$$

Where g is the acceleration of gravity, β is the coefficient of thermal expansion, ΔT-w is the temperature difference of the water in the tank, v_w is the kinematic viscosity and a_w is the thermal diffusion coefficient.

The heat transfer coefficient $a_{c,r}$ on the refrigerant side is determined by :

$$\alpha_{c,r} = \frac{Nu_{c,r} \cdot \lambda_{c,r}}{d_{c,i}} \tag{3-19}$$

Heat transfer on the refrigerant side is divided into three zones as shown in Figure 3.6. The convective heat transfer coefficient $a_{c,r}$ is different in each flow zone and can be divided as followsa$_{sh,c}$, $a_{tp,c}$ and $a_{sc,c}$.

The convective heat transfer coefficient in single-phase operation,

$$Nu_{c,r} = 0.023Re_{c,r}^{0.8}Pr_{c,r}^{0.4}\delta^{0.1} \tag{3-20}$$

$$\alpha_{sh,c} = \frac{Nu_{sh,c,r}\lambda_{sh,c,r}}{d_{c,i}} \tag{3-21}$$

$$\alpha_{sc,c} = \frac{Nu_{sc,c,r}\lambda_{sc,c,r}}{d_{c,i}} \tag{3-22}$$

In the two-phase regime, the heat transfer coefficient ($a_{tp,c}$) is determined by the following correlation:

$$\alpha_{tp,c} = 1.3\alpha_{sc,c}\left[(1-x)^{0.8} + \frac{3.8x^{0.76}(1-x)^{0.04}}{Pr_{tp,c,r}^{0.38}}\right] \tag{3-23}$$

Where a_{sc} is the heat transfer coefficient at the condenser (undercooled liquid), x is the vapour content of the two-phase mixture and $Pr_{tp,c,r}$ is the Prandtl number in a liquid-vapour mixture.

2.6.1.4.2 Model assumptions

A number assumptions have been made to make this model easier to implement:

- The three-dimensional geometry of the condenser immersed in a water tank could be simplified into a two-dimensional axisymmetric model as shown in Figure 3.7.
- The helical shape of the condenser has been simplified to several circles in a plane of symmetry.

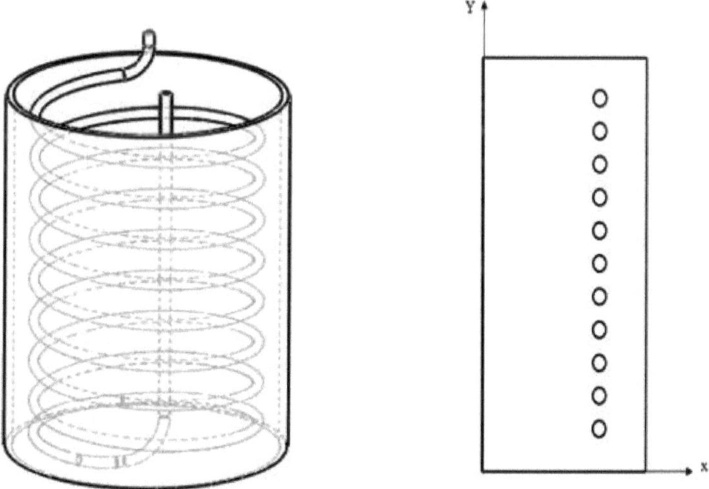

(a) 3D design of a water with a fully immersed condenser

(b) 2D design of a water with a fully condenser

Figure 3. 7: Presentation of the geometry

2.6.1.4.3 Theoretical model of condenser length

In terms of sizing a heat exchanger, a numerical study is taken into account because it is quicker and more efficient for identifying the various physical phenomena and heat transfers. It also enables us to minimise the cost of experimental testing. For the modelling, we followed the method detailed in our previous paper **[54].**

Under these conditions, the cooling capacity is defined by :

$$Q_r = \dot{m}_w \cdot Cp_w \cdot \Delta T \tag{3-24}$$

With :

riiw: The mass flow rate of water $^{kg}/_{s}$

Cp_w: The specific heat of water $^{kJ}/_{kg \cdot k}$

ΔT: The difference in water temperature in k

Evaporation energy is calculated using the following formula:

$$Qe = h_2 - h_3 \quad en \ ^{kJ}/_{Kg} \tag{3-25}$$

The mass flow rate of the R134a refrigerant is defined by the following expression :

$$\dot{m}_r = {}^{P_{cr}}/_{Q} \tag{3-26}$$

The following expression used to determine the condenser cooling water flow rate rii$_{(cw)}$:

$$\dot{m}_{cw} = \frac{\dot{m}_r \cdot (h_2 - h_3)}{Cp_{cw} \cdot (t_{cw2} - t_{cw1})} \tag{3-27}$$

Where: mir: Refrigerant mass flow rate enk $^{kg}/_{s}$

h_2 and h_3: The enthalpy of mass at the condenser inlet and outlet in $^{kJ}/_{kg}$

$Cp_{(cw)}$: Specific heat of water in $^{kJ}/_{kg \cdot k}$

t_{cw1} and t_{cw2} : The temperature of the cooling water at the condenser inlet and outlet in kelvin (k).

The condenser cooling water volume flow rate $V_{(cw)}$ at $^{m^3}/_{s}$ is written as follows:

$$\dot{V}_{cw} = \frac{\dot{m}_{cw}}{\rho_{cw}} \tag{3-28}$$

With

mcw: Condenser cooling water mass flow rate in $^{kg}/_{s}$

pcw: Density of water in $kg/_{m3}$

The speed of the cooling water in m/s is given by the following expression:

$$V_{cw} = \frac{4 \cdot \dot{V}_{cw}}{(3.14 * (do^2 - di^2))} \tag{3-29}$$

Where: d_0: External diameter of the condenser tube in m.

d_i : Internal diameter of the condenser tube in m.

Knowing that the temperature of the cooling water at the condenser inlet is given by $T_{(cw1)}$ in (°C) and that the temperature of the cooling water at the condenser outlet is given by T_{cw2} in (°C), we can deduce the average temperature :

$$T_{moy} = \frac{t_{cw1}+t_{cw2}}{2} \quad (3\text{-}30)$$

For this average temperature T_{avg}, the thermo-physical properties of the water are determined using the Engineering Equation Solver (EES) software.

Determination of the heat transfer coefficient ($\grave{a}_0$) **of water:**

The Reynolds number is defined on the basis of the thermo-physical properties of water. For the flow outside the tube we have :

$$Re = \frac{\rho VD}{\mu} \quad (3\text{-}31)$$

With: ρ density of the fluid, V characteristic velocity of the fluid (velocity at a distance from the wall), D diameter of the tube, μ dynamic viscosity.

To calculate the Prandtl number of the external flow of the tube, we use the following formula:

$$Pr = \frac{\mu C_p}{\lambda} \quad (3\text{-}32)$$

With

μ the dynamic viscosity, C_p the mass heat capacity at constant pressure, λ the thermal conductivity.

To calculate the Nusselt number, we use the Dittus-Boelter equation:

$$Nu = 0.023 * Re^{0.8} * Pr^{0.3} = \left(\frac{h_o D}{\lambda}\right) \quad (3\text{-}33)$$

With :

Re: Reynolds number

Pr: Prandtl number

The heat transfer coefficient is written as :

$$h_o = \frac{Nu \cdot \lambda}{D} \quad (3\text{-}34)$$

With :

Nu: Nusselt number.

λ: Thermal conductivity of water in $W/_{m.k}$

D: External diameter of the pipe in m

Determination of the heat transfer coefficient (h_i) **of refrigerant R134a:**

Knowing that *Re* is the Reynolds number of the flow inside the tube and given by the following expression :

$$Re = \frac{\rho VD}{\mu} \quad (3\text{-}35)$$

To calculate the Prandtl number for the internal flow of the refrigerant, we use the following formula:

$$Pr = \frac{\mu C_p}{\lambda} \quad (3\text{-}36)$$

The equation proposed by Dittus-Boelter is used to calculate the Nusselt number:

$$Nu = 0.023 * Re^{0.8} * Pr^{0.3} = \frac{hD}{\lambda} \quad (3\text{-}37)$$

$$h_i = \frac{Nu \cdot \lambda}{D} \quad (3\text{-}38)$$

With :

h: Heat transfer coefficient on the refrigerant side in $W/_{m^2\,K}$

λ: Thermal conductivity of the refrigerant in $W/_{m.k}$

D: Internal diameter of the pipe in m

μ: Dynamic viscosity of the refrigerant in Pa. s

Cp: Specific heat of the refrigerant in $J/_{Kg.K}$

ρ : Density of the refrigerant in $Kg/_{m^3}$

V: Velocity of the refrigerant inside the tube in $m/_s$

To evaluate the performance of a heat exchanger under steady-state conditions, the logarithmic mean temperature difference DTLM can be written as follows:

$$DTLM = \frac{\Delta\theta_1 - \Delta\theta_2}{Ln(\frac{\Delta\theta_1}{\Delta\theta_2})} \qquad (3\text{-}39)$$

Temperature differences are :

$$\Delta\theta_1 = |T_{f.s} - T_{eau.e}| \tag{3-40}$$

$$\Delta\theta_2 = |T_{f.e} - T_{eau.s}| \tag{3-41}$$

The total heat transfer coefficient *(U0)* can be written as follows:

$$U_0 = \frac{1}{\frac{1}{h_i} + \frac{1}{h_o} + \frac{1}{\frac{e}{\lambda}}} \tag{3-42}$$

With :
h_i : Thermal convection coefficient on the refrigerant side.
h_o: Water-side thermal convection coefficient.
e: wall thickness.
λ: thermal conductivity.
Consequently, the length of the condenser can be written as :

$$Q = U_0 \cdot A \cdot \theta_m \tag{3-43}$$

$$Q = U_0 . (3.14 \cdot D \cdot L) \cdot (\Delta\theta_1 - \Delta\theta_2) \tag{3-44}$$

$$L = \frac{Q}{U_0 \cdot 3.14 \cdot D \cdot (\Delta\theta_1 - \Delta\theta_2)} \tag{3-45}$$

With :
Q: Power exchanged between the water and the refrigerant at the condenser in W
U_0: Overall heat transfer coefficient in W. $m^{(-2)}k^{-1}$
D: External diameter of the pipe in m
ΔΘ1 and ΔΘ2: Temperature differences between the two fluids (refrigerant and external fluid expressed in°C.

2.6.1.5 Modelling the water

For a condenser cooled by natural water convection, the refrigerant circulating inside the tube is cooled by the water circulating outside the tube. Therefore, to facilitate modelling, a large number and variety of assumptions that are available in recent work are taken into account. Note that the assumptions chosen in our mathematical model are selected for their ability to reproduce the experimental results. To simplify this problem, the model has been assumed as shown in Figure 3.8.

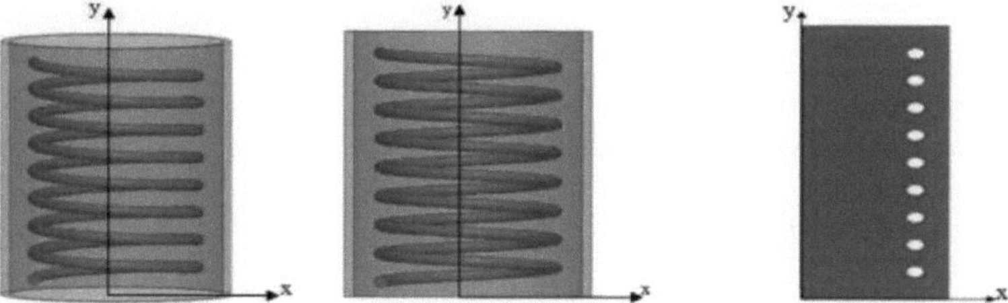

(a) 3D reservoir design (b) 2D reservoir design (c) 2D axisymmetric design
Figure 3. 8: Simplified model of the water with a fully immersed condenser
[50].

Model assumptions

The water reservoir model is developed on the basis of the following assumptions:

- The design of the water in 3D is simplified in 2D Axisymmetric.
- The volume of hot water in the tank is preserved, i.e. no hot water is consumed during the heating period.
- The size of the hot water storage tank is chosen according to the minimum requirement for supplying hot water.
- Fluid flow is assumed to be two-dimensional, unsteady and laminar.
- The Boussinesq hypothesis is used to model fluid flow with natural convection.
- The ends of the tank are adiabatic, i.e. the tank is well insulated and heat loss from the tank to the environment is neglected (no heat loss from the tank wall to the atmosphere).

2.6.2 Method of coupling the refrigerator to the water heater

To pair a domestic refrigerator with a water heater, the following steps were taken:

- **Step 1:** Define an initial heat flux $q(t)_0$ from the experimental tests.
- **Stage 2 :** Run the CFD code in transient mode for the entire heating time.
- **Stage 3 :** Obtain the velocity profile v(t) and temperature profile T(t) of the water.
- **Stage 4 :** A new heat flux $q(t)_i$is determined using the MATLAB programme.
- **Stage 5 :** Obtain a heat flux $q(t)_i$ that varies as a function of time over the entire heating time.

$$q(t)_i = A.t^3 + Bt^2 + C.t + D \tag{3-46}$$

- **Step 6:** A new heat fluxq(t)¿ found is then uploaded into the CFD code as a boundary condition for the helicoidal condenser in order to obtain the new water-side information.
- **Step 7:** Obtain the water temperature profile, water velocity, heat transfer and system coefficient of performance.

2.6.3 Determining the quantity of water to be heated by the condenser

The following method was used to optimise the quantity of water to be heated daily by using waste heat from the condenser:

In the heat exchanger (condenser), the heat supplied by the refrigerant is gained by the water.

$$Q_c = Q_{eau} \tag{3-47}$$

Knowing that :

$$COP \cdot P_{comp} \cdot t \cdot \eta = m_{eau} \cdot Cp \cdot \Delta T \tag{3-48}$$

With :

COP: Condenser coefficient of performance

P_c : Compressor power

t: Compressor operating time

η: Heat storage efficiency.

m_{water} : Quantity of water to be heated daily in kg

Cp: Specific heat of water $J/_{kg.°C}$

ΔT: Change in water temperature, from initial to final state, in the condenser in °C For the determination of the water heat storage efficiency, the following formula is used:

$$\eta = Q_s/Q_a \tag{3-49}$$

With :

Q_s: Quantity of heat stored by the water

Q_a: Quantity of total heat received by the water

So the two quantities of heat can be written as :

$$Q_a = P_{comp} \cdot COP \cdot t \tag{3-50}$$

$$Q_s = Q_{abs} - Q_p \tag{3-51}$$

With :

Q_p: Amount of heat lost by the water

P_{comp}: Compressor power

The amount of heat lost by the water in the condenser can expressed as :

$$Q_p = U \cdot \Delta T \cdot t \tag{3-52}$$

With :

ΔT: Difference in water temperature in the tank.

t : Compressor operating time

U : Heat loss coefficient

To determine the heat loss coefficient, we use the following equation

$$U = \frac{\rho \cdot Cp \cdot V}{\Delta t} \cdot \ln\left(\frac{T_i - T_{am}}{T_f - T_{am}}\right) \tag{3-53}$$

With

ρ: Density of water in Kg/m^3

Cp: Specific heat of water in $J/Kg.°C$

V: Volume of water in m^3

Δt : Compressor operating period

T_{am}: Ambient temperature in °C

T_i and T_f: Initial and final water temperatures in the condenser respectively in °C

The result is :

$$\eta = \frac{P_c\, COP\, t - U\, \Delta T\, t}{P_c\, COP\, t} \tag{3-54}$$

The variation in water temperature is defined by :

$$\Delta T = \frac{P_{comp}\, COP\, t}{U\, t + m_{eau}\, Cp} \tag{3-55}$$

Therefore, to determine the mass of water to be heated daily, we use the following formula:

$$M_{eau} = \frac{P_{comp}\, COP\, t}{\Delta T \cdot \left(\frac{Cp}{\Delta t} \ln\left(\frac{T_i - T_{amb}}{T_f - T_{amb}}\right) t + Cp\right)} \tag{3-56}$$

2.7 Numerical method

In this study, we simulated a theoretical model using the ANSYS Fluent code to take into account the geometry of the condenser, the water tank and, above all, the heat transfer between the refrigerant in the condenser and the water in the tank. Using a mathematical model, the variation in temperature and water velocity are determined. The heat transfer and coefficient of performance of the system can also be studied. To study the thermal behaviour of the condenser and its coupling with a water tank, we used the ANSYS Fluent commercial code, which is well suited to solving the conservation equations under variable conditions and to taking into account the coupling within the domestic refrigerator studied.

2.7.1 Governing equations

The convective flow of the refrigerant inside the condenser and of the water in the storage tank is modelled on the basis of the principles of conservation of mass, momentum and energy. Based on the assumptions adopted, the governing equations of the model can be written, in Cartesian form, as follows:

Continuity equation

$$\frac{\partial U_x}{\partial x}+\frac{\partial U_y}{\partial y}=0 \qquad (3\text{-}57)$$

With *u* and *v* representing the components of the fluid velocity in the *x* and *y* directions respectively

The momentum equation is written along the axis (ox) :

$$\frac{\partial U_x}{\partial t}+U_x\frac{\partial U_x}{\partial x}+U_y\frac{\partial U_x}{\partial y}=-\frac{1}{\rho wo}\frac{\partial P}{\partial x}+\upsilon_w\left[\frac{\partial^2 U_x}{\partial x^2}+\frac{\partial^2 U_x}{\partial y^2}\right] \qquad (3\text{-}58)$$

Along the (oy) axis, we write :

$$\frac{\partial U_y}{\partial t}+U_x\frac{\partial U_y}{\partial x}+U_y\frac{\partial U_y}{\partial y}=-\frac{1}{\rho wo}\frac{\partial P}{\partial y}+\upsilon_w\left[\frac{\partial^2 U_y}{\partial x^2}+\frac{\partial^2 U_y}{\partial y^2}\right]-g\beta(T_w-T_{wo}) \qquad (3\text{-}59)$$

The conservation of energy equation can be written as

$$\frac{\partial T}{\partial t}+U_x\frac{\partial T}{\partial x}+U_y\frac{\partial T}{\partial y}=\lambda_w\left[\frac{\partial^2 T}{\partial x^2}+\frac{\partial^2 T}{\partial y^2}\right] \qquad (3\text{-}60)$$

2.7.2 Initial and boundary conditions

- Initial conditions

As shown in Figure 3.9, the condenser is in helical form, fully immersed in the water during experimental study. In the CFD modelling, the condenser is considered as small circles in a plane. The heat flux given off by the condenser is determined from the experimental results and added to the CFD model as initial conditions. The initial water temperature is of the order of 20°C and the ambient temperature is 21°C. An acceleration component (gravity acceleration g=9.81 m.s^{-2}) is also defined.

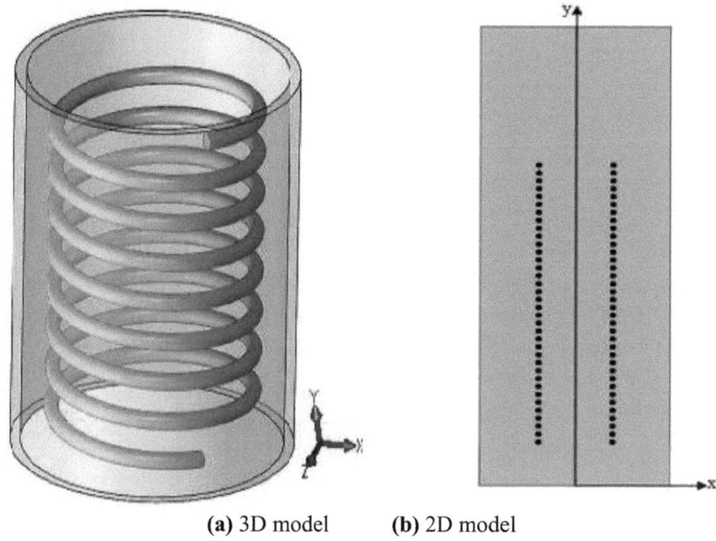

(a) 3D model **(b)** 2D model

Figure 3. 9: Water tank with a fully immersed helicoïdal condenser **[55].**

At the start of the simulation, the velocity of the water inside the tank, the density, the pressure and the temperature are considered to be constant. The initial model parameters are those shown in Figure 3.10.

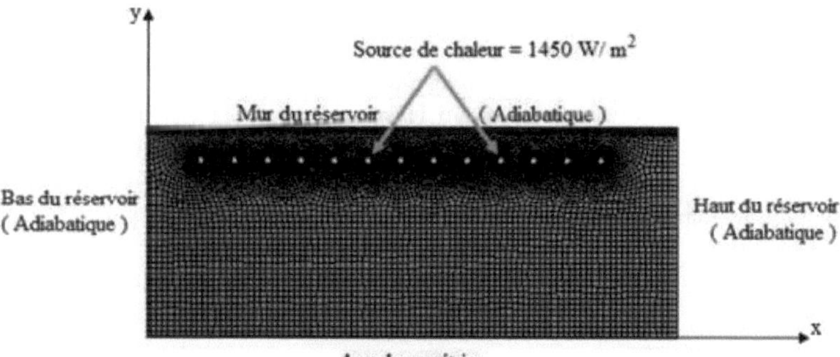

Figure 3. 10: Boundary conditions in a 2D configuration taken into account during simulation

- Boundary conditions

A simulation based on the finite volume technique was selected for the modelling. The latter is carried out for two conditions, time-varying heat flux calculated by implementing Fluent's UDF (User Defined Function) and constant flux. The "SIMPLE" method is used for pressure-velocity coupling with a second-order spatial discretization for energy and momentum. The PRESTO method is used to treat the pressure. The sub-relaxation factors for the energy, pressure and momentum equations are 1, 0.3 and 0.7 respectively. The results are analysed when the convergence criteria (set at 10^{-6}) on the residuals of the solved equations are reached. On the "Botton, Top and Wall" edge, an "adiabatic" type condition is imposed. On the axis of symmetry of the reservoir, an "Axis" type condition is imposed as shown in Table 3.1.

Table 3.1 Basic parameters used in the simulations

Low	Adiabatic
Top	Adiabatic
Wall	Adiabatic
Axisymmetry	Axis
Condenser	Variable heat flux + UDF/ constant heat flux

The thermo-physical properties of water are determined using Engineering Equation Solver (EES) software based on the relationship between temperature and pressure obtained from the experiment that was carried out. Table 3.2 highlights the main properties used.

Table 3.2: Physical properties of water used in CFD simulation

Water properties	Symbol	Value	Unit	Method
Density	ρ	994	kg/m^3	Boussinesq
Specific heat	Cp	4183	J/(kg.K)	Constant
Thermal conductivity	λ	0.6107	W/(m. K)	Constant
Dynamic viscosity	μ	0.0007196	kg/(m. s)	Constant
Coefficient of thermal expansion	β	0.000347	1/K	Constant

2.7.3 CFD modelling (Computational Fluid Dynamics)

2.7.3.1 Creating geometry

Figure 3.9 shows the geometric layout of the prototype tank with a fully submerged helical condenser. In order to simplify the model, certain assumptions are made. The height of the helical condenser in the vertical direction of the tank is small and could be neglected. The condenser turns are assumed to be small circles aligned in the plane of symmetry. Also, as we observe, the 3D geometry of the water tank could be simplified to an axisymmetric 2D model. Therefore, in our case, we will work in 2D axisymmetric in the XY plane to create the geometry. The tank geometry was created using ANSYS Design Modeler. Figure 3.11 shows the dimensions chosen to create the geometry in question.

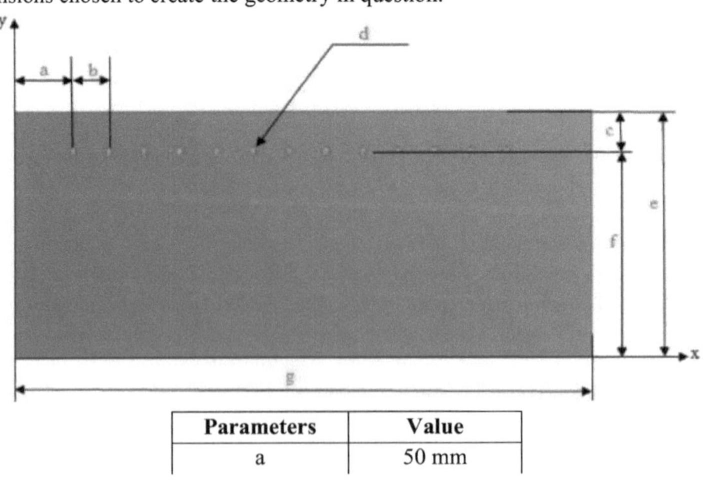

Parameters	Value
a	50 mm

b	30 mm
c	30 mm
d	6 mm
e	190 mm
f	160 mm
g	480 mm

Figure 3. 11. Creating the geometry using Design-Modeler

2.7.3.2 CFD mesh

In this section, we present a validation of the mathematical model by analysing the independence of the mesh size and time step, and by comparing our numerical results with experimental results.

Two-dimensional mesh

The two-dimensional mesh of the reservoir geometry is produced using ANSYS Fluent software version R16.2. It contains 3000 nodes with a non-structural mesh with triangular and quadrilateral volumes belonging to the domain. This mesh can guarantee better convergence and higher resolution. The influence of the mesh on the simulation is systematically checked at the start of the simulations. The results presented are independent of the mesh. Also, the influence of the time step on the calculation is tested for time steps equal to 0.1 s, 0.5 s, 1 s and 10 s. In addition, all the Fluent calculations are performed in double precision. Figure 3.12 gives an overview of the mesh used.

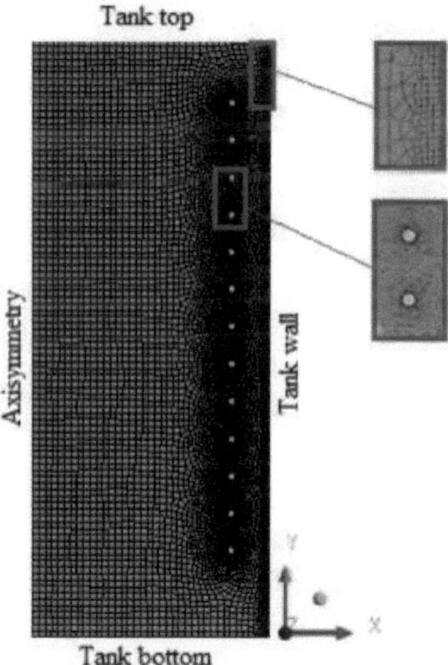

Figure 3. 12. Mesh of the geometry used

Effect of the mesh Before using the numerical model, it is very important to analyse the effect of the mesh. To do this, we studied four mesh grids corresponding to a total number of 4162, 24202, 58989 and 122005 nodes. The variation in water temperature in the reservoir is chosen for comparison (Figure 3.13). It can be seen that the differences obtained for the four grids studied are relatively small. We then adopted the grid with 24202 nodes for our calculations, which gives a good compromise between accuracy and calculation time. In addition, several time steps varying between 0.1 and 10 s were tested on the adopted grid. Under these conditions, we chose the value At=1 s with a maximum iteration of around 200 in each time step (Figure 3.13).

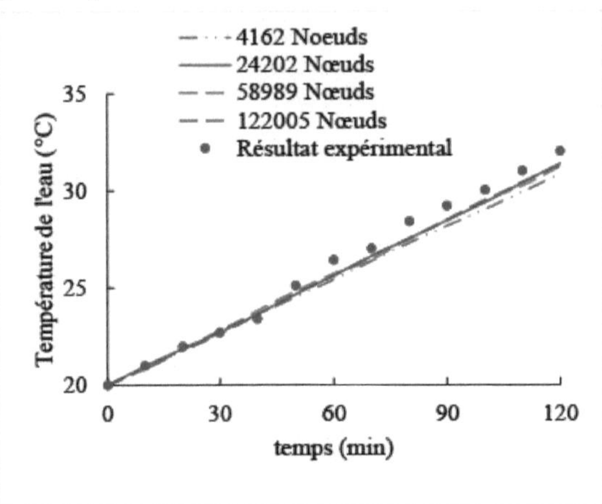

Figure 3. 13. Temporal variation in mean water temperature for different two-dimensional meshes.

Effect of time step (At)

Before starting the simulations, we wanted to know the effect of the time step At on our calculations by varying it according to the average water temperature during the heating phase. To check the effect of the time step on our results, we compared the curves of variations in water temperatures during the heating phase with the recovery of waste heat from the condenser by varying the time steps, for values of At = 0.1 s , At = 0.5 s , At = 1 s and At = 10 s.

The results presented in Figure 3.14 show that the effect of the time step on the heating process becomes very small when the number of steps decreases from 10 s to 0.1 s. For this reason, we opted for At = 1 s, in order to achieve a compromise between accuracy and calculation time.

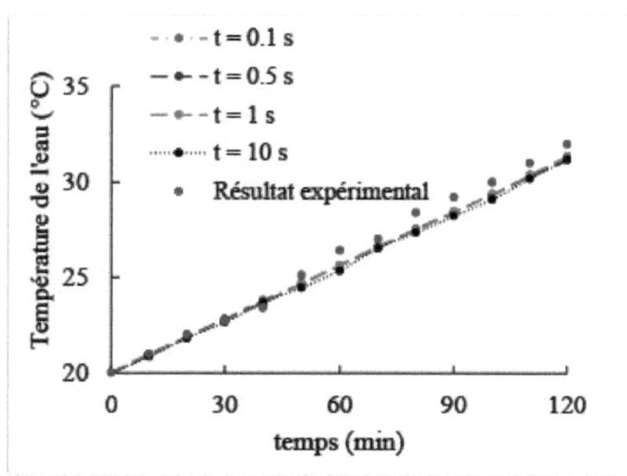

Figure 3. 14. Time variation of the mean water temperature for different values of Δt

2.7.3.3 Post Processing

Speed/pressure coupling (SIMPLE algorithm)

In order to solve the Navier-Stokes equation which describes fluid motion, it is necessary to know the pressure field. If this field is unknown, a pressure equation must be established. This specificity of the equations makes it necessary to use a pressure-velocity coupling scheme. Consequently, in the present work we use the SIMPLE (Semi Implicit Method for Pressure Linked Equations) algorithm proposed by **Dai et al [56]** and **Wenzli et al [57].**

Discretisation scheme

The gradient of the variables on the mesh geometry is calculated using the "Least Squares Cell Based" method. The pressure interpolation method at the cell faces is carried out using the "PRESTO" scheme. In the present study, the "Second Order Upwind" scheme is applied to resolve the energy and takes into account the direction of fluid flow.

Implementing UDF

The commercial software ANSYS Fluent is used to develop a 2D axisymmetric model of the water tank with a helicoidal condenser. The variation of the heat flux given off by the condenser as a function of time is calculated by a Fluent function (UDF: User Defined Function). At any instant, the heat flow can therefore be considered to be variable as a function of time, decreasing as the water temperature increases.

Consequently, the heat flux is a new variable managed by the UDF. At each time step, the flux varies. ANSYS FLUENT software is used to run simulations. One of the major advantages of this code is that it devotes a large part to modelling the operation and coupling of the refrigerator to the water heater. A UDF (User Defined Function) file in the form of a subroutine written in C language is produced and included in the calculations. **Convergence conditions**

The calculation over a time step is considered as converged when the normalised residuals of the solved equations have reached the value of $10^{(-6)}$ for the continuity equation; the momentum equation and the energy equation. These values are set by default during the calculation and are the recommended values.

Under-relaxation

Sub-relaxation of the values obtained at each iteration is regularly used in non-linear problems to avoid divergence in the calculation process. This allows us to control and reduce the change for each iteration. To improve convergence stability, sub-relaxation factors of 0.3, 1, 1, 0.7 and 1 s are considered for pressure, density, volume force, momentum and energy respectively. **Convergence**

The convergence criterion is a specific condition for the residuals that define the convergence of a solution. It is therefore used to stop the iterative process when it is satisfied. In our study, convergence is assured when the maximum of the residual of all variables reaches the value of 10^{-6}.

2.8 Model validation

The best way to validate a numerical model and ensure its functionality and reliability is to carry out a comparative study between simulation and experimental results. In our case, we begin by validating our numerical results with the experimental and numerical work published by **Dai et al [8]** and **Wang [58]**. In the same line of research, a numerical study is carried out on another geometry. To verify the mathematical model, a comparison is made between the numerical and experimental results. Once the model has been validated, a parametric study can be carried out to optimise the thermal behaviour of a helical condenser fully immersed in a water tank.

2.8.1 Calculation according to D.D Wang's conditions [58].

Our numerical results are compared with the experimental data published by **[58]**. In fact, we carried out simulations in ANSYS FLUENT and compared them with those available in similar studies, such as those by **Wang [58]**. Figure 3.15 shows the profile of mean water temperature as a function of reservoir height. Figure 3.15 shows the comparison between the different results. Overall, this comparison shows good agreement with those available in the literature, with an error of 10.96%.

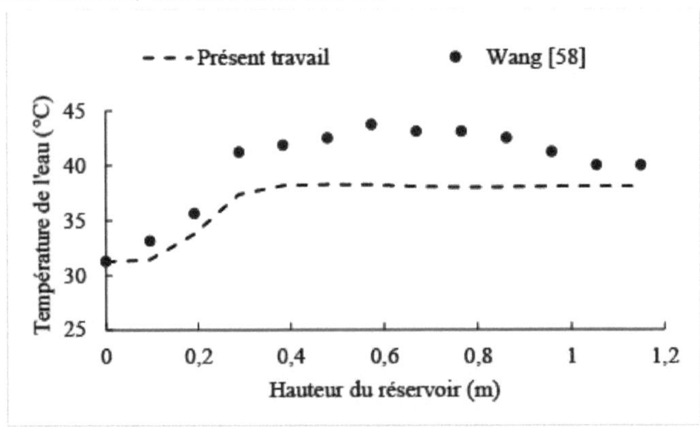

Tank height (m)

Figure 3. 15. Comparison of the variation in mean water temperature with the work of **Wang**'s work **[58]**.

2.8.2 Calculation according to the conditions of Dai et al [8].

Figure 3.16 shows that the gap between the results of the proposed model and those available in the literature is narrow. The average relative error is equal to 3%. Therefore, these deviations allow us to correctly predict the effect of some parameters on the performance of the chiller coupled with a domestic water heating unit.

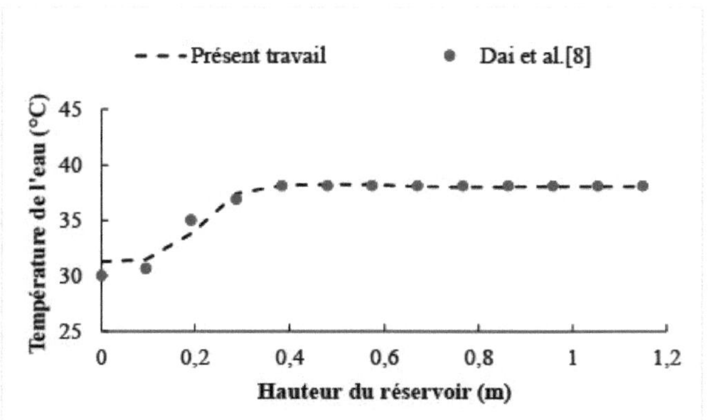

Figure 3. 16. Comparison of the variation in mean water temperature with the work of **Dai et al [8].**

Figure 3.17 shows the variation in water temperature as a function of tank height. These results show good agreement with the experimental results obtained by **Wang [58]** and the numerical results obtained by **Dai et al [8]** with maximum observed errors of the order of 10.96%. If we compare the numerical and experimental results of our numerical study, we can see that these errors are smaller. Consequently, these deviations allow us to correctly predict the thermal behaviour of the coupled system.

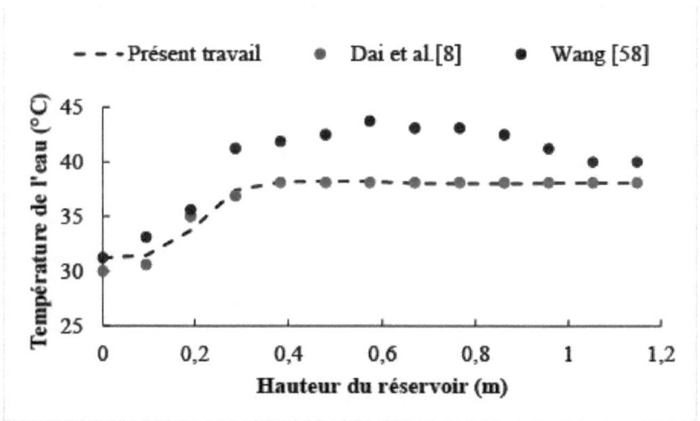

Figure 3. 17. Comparison of the variation of the mean water temperature in the reservoir with the work of **Dai et al [8]** and **Wang [58].**

2.8.3 Calculation according to experimental conditions

Figure 3.18 shows the variation of the water temperature in the tank as a function of time. Our numerical results are compared with experimental results from the Energy, Water, Environment and Processes research laboratory at the National Engineering in Gabès. Our numerical results show good agreement with the experimental results, with average relative errors of the order of 7.67%.

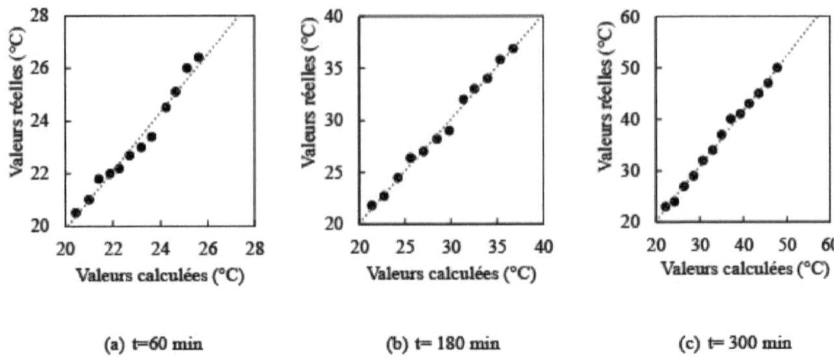

Figure 3. 18. Water temperature profile as a function of time

3 Conclusion

In this chapter, we have presented the simulation stages using the ANSYS Fluent calculation code. We cited the method of resolution and the operating principle. We have described the steps taken to create a numerical model for calculating the length of the condenser immersed in a water tank. We also presented a model for calculating the quantity of water to be heated daily by the refrigerator coupled to the water heater. Next, we described a mathematical model that enabled us to study and analyse the numerical results of a coupled system by modifying geometric parameters. In addition, we evaluated the numerical model used to describe the effect of the geometry of the helical condenser on thermal behaviour, the results of which will be discussed in the next chapter. Finally, we compared our numerical results with other results from the literature and our experimental results developed in our laboratory.

CHAPTER 4

Results and discussion

1 Introduction

This final chapter looks at the coupling of a domestic refrigerator to a water heater and an air conditioner for seawater desalination. The contribution will be made through both theoretical and experimental studies. To this end, we have divided this chapter into two main parts: The first part analyses the performance of these refrigeration machines in order to confirm the importance of thermal rejection for the production of domestic hot water. This part presents the numerical study of the main parameters influencing the operating characteristics of a refrigerator coupled to the water heater and the thermal behaviour of the helicoïdal condenser fully immersed in a thermally insulated water . We also present our numerical results relating to the study of the effect of the condenser geometry. We then study the various parameters for improving the performance of the refrigerator. The second part is devoted to the presentation of the experimental results obtained with reference to the brackish water desalination technique using the thermal discharge from the air-conditioner condenser. In this part, we are interested in doubling the overall energy performance of these refrigeration machines.

2 Production of domestic hot water

In this section, we present our numerical and experimental results on the phenomenon of heat transfer through the use of refrigerator waste heat. The evolution of velocities and temperatures are given to represent the importance of thermal rejection for the production of domestic hot water in the industrial and residential sectors. Some numerical results on the effect of condenser geometry on the thermal performance of a domestic refrigerator coupled with a domestic hot water heating unit are also reported. The numerical results of heat transfer by natural convection between the condenser and the water will be represented by the evolution of velocities, temperatures and the distribution of the pressure field, the temperature field and the water velocity field for the different conditions. Firstly, the effect of geometric parameters on the heat transfer phenomenon is presented. Secondly, the choice of condenser design and construction, which involves improving the performance of the refrigeration machine used to produce domestic hot water. This numerical study will make it possible to choose the best geometry for the most economically and energetically efficient condenser.

2.1 Experimental and numerical study of a domestic refrigerator coupled to a water heater

In order to fully understand the importance of the heat rejected by the domestic refrigerator condenser, we begin by analysing the experimental results, which show the evolution of the temperature as a function of time, the temperatures of the water to be heated, the evaporator and the ambient air. The experimental study was carried out under environmental conditions defined by an ambient temperature of 20°C. A cylindrical tank containing 50 litres of water is heated from 20°C to 50°C over a total operating period of t=300 min. The temperature trends are shown in Figure 4.1. From these results, it can be seen that for the heating time equal to t=300 min, the water temperature increases progressively but the evaporator temperature varies between -13°C and -17°C. The results obtained show that the evaporator temperature is not affected by the increase in water temperature in the tank. These results therefore confirm that the domestic refrigerator can be used to produce hot water without affecting its main refrigeration function.

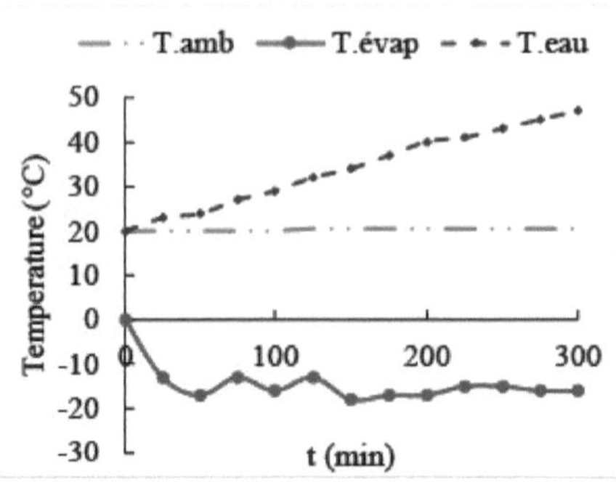

Figure 4.1: Temperature variation as a function of heating time

Figure 4.2 shows the variation in water temperature and system performance as a function of heating time. From these results, we can see that there is a relationship between the refrigerator's coefficient of performance and the water temperature. As the temperature of the water in the tank increases, we observe a significant decrease in the COP. In fact, when the water temperature is lower, the COP is higher. These results show that the COP around 4.8 at t=25 min and decreases as the water temperature inside the tank increases. It can also be seen that as the water temperature increases, the COP gradually decreases. In fact, as the water temperature increases, the COP decreases more rapidly. When the water temperature is **22.22** °C, the COP reaches 4.8. When the water temperature rises by **2.04°C**, the COP drops to 4.3. However, when the water temperature rises by 13°C, the COP decreases to 3. Thus, the coefficient of performance (COP) of the domestic refrigerator coupled with a domestic hot water production unit decreases during the heating process. The results obtained showed that when the water temperature rises, the efficiency of the system decreases considerably. As a result, the efficiency of the compressor is affected by the increase in the temperature of the water in the tank. This is because when the water temperature increases, the amount of energy used also increases **[59].** Furthermore, this justification indicates the effect of the water temperature inside the tank on the system's performance.

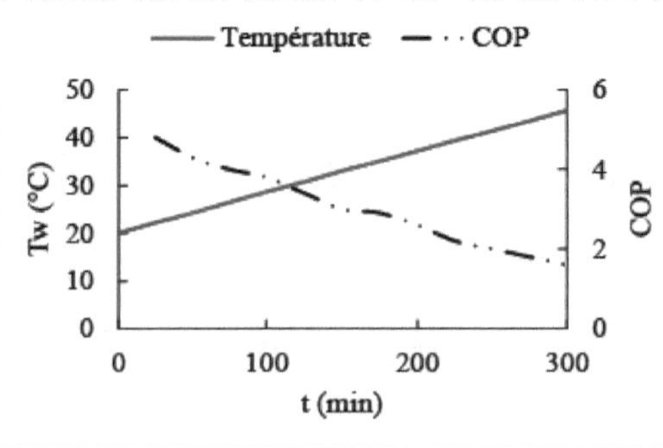

Figure 4. 2: Variation in water temperature and refrigerator performance as a function of time

2.1.1 Temperature profile

Figure 4.3 shows the evolution of water temperature as a function of tank height for different heating times defined by t= 60 min, t= 180 min and t= 300 min. From the calculations carried out, it can be seen that the water temperature increases along the height of the tank for three different heating periods. For the first heating period equal to t=60 min, we can see that the average water temperature is equal to T= 25.50°C while the minimum and maximum temperatures over the height of the tank are equal to T=20.46°C and T= 26.29°C respectively. During this heating period (t=60 min), the change in water temperature is too small in the lower part of the water tank, leading to the creation of what is known as stratification. This is produced by a buoyancy-driven flow. As for the upper and middle parts of the reservoir, the change is clear. These results show that the water temperature increases with the height of the tank and becomes stable from the position defined by Ht=0.15m. When the heat given off by the refrigerator condenser is recovered to heat the water in the tank for t=180 min, the average temperature of the water increases and becomes equal to 36.6°C. However, in the vertical direction of the tank, the difference in maximum temperature is equal to T=13.25 °C. As the heating time increases, the water temperature rises with the height of the tank, with behaviour varying according to the heating period. We can see that for a refrigerator operating time equal to t=300 min, the water is heated from T=20 °C (initial state of the water) to T=50.73 °C (final state of the water). When these results were analysed, the mean, minimum and maximum temperatures in the vertical direction of the tank were equal to T=48 °C, T=32.7 °C and T=50.7 °C respectively. Consequently, after about 60 minutes of heating, the water temperature is too low to be used. The water in the tank would then be heated continuously for 240 minutes. During this heating period, the water is heated to T=50.7 °C. This value is also very useful for homeowners who use hot water.

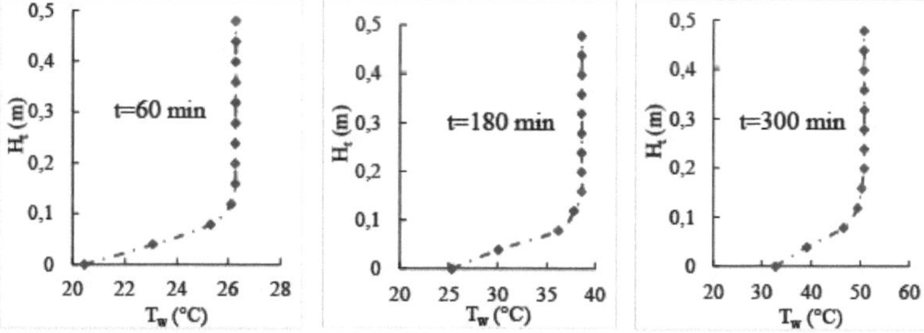

Figure 4.3: Variation in water temperature as a function of tank height

2.1.2 Speed profile

Figure 4.4 shows the variation in water velocity as a function of tank height for three different heating periods defined by t=60 min, t=180 min and t=300 min. Using the numerical calculation code, the water velocity profiles at different heating times are compared and analysed. It can be seen that for t=60 min, the maximum water velocity is equal to V=0.013 $m.s^{-1}$ and appears in the middle part of the tank (Ht<0.3 m). This velocity distribution is almost obvious in this section. At t= 180 min, the water velocity reaches its maximum in the middle part of the tank. For the heating time t=300 min, a low velocity still exists in the lower part of the tank. As a result, as the heating time increases at t=60 min and t=300 min, the hot water continuously flows upwards and accumulates in the upper part of the tank, leading to the formation of a higher velocity zone.

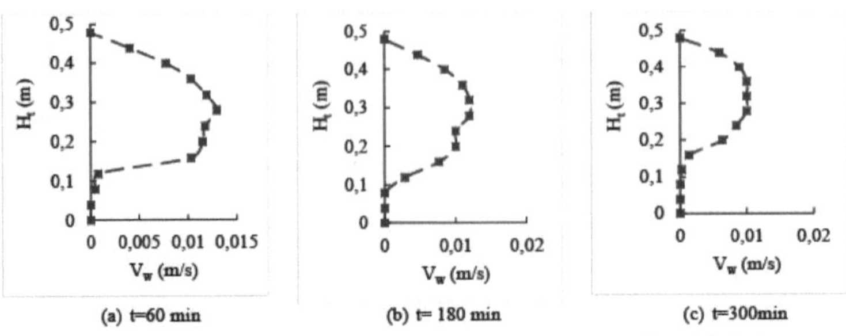

Figure 4.4: Variation in water velocity as a function of tank height

2.1.3 Heat transfer coefficient profile

Figure 4.5 shows the variation of the convective transfer coefficient as a function of the heating time, which is defined as t=60 min, t=180 min and t=300 min. According to the simulation results obtained, there is a decrease in the heat transfer coefficient during the heating period. This can be explained by the fact that the increase in the temperature of the water in the tank leads to a decrease in the convective exchange coefficient. To examine the influence of the water temperature on the convective exchange coefficient, we can see that for t=60 min, the heat transfer coefficient decreases from 727.26 $W.m^{-2}.K^{-1}$ to 253.6 $W.m^{-2}.K^{-1}$. At t=180min, it decreases from 558.23 $W.m^{-2}.K^{-1}$ to 102 $W.m^{-2}.K^{-1}$. By comparing the two profiles, it can be seen that during the heating process, the heat transfer coefficient decreases with increasing water temperature. The greatest difference approximately 159.95 $W.m^{-2}.K^{-1}$. Consequently, the decrease in the heat transfer coefficient is due to the increase in the temperature of the water in the tank, which leads to a reduction in the temperature difference at the condenser inlet and outlet. At t= 300 min, the maximum and minimum values of the condenser heat transfer coefficient are respectively equal to 452.27 $W.m^{-2}.K^{-1}$ and 68.64 $W.m^{-2}.K^{-1}$. Comparing these results over the three periods considered, the heat transfer coefficient of the condenser at t= 60 min is higher than in the other two times equal to t=180 min and t=300 min. This can be explained by the increase in water temperature, which is responsible for the decrease in the heat transfer coefficient. This study shows that as the water temperature increases, the convective exchange coefficient decreases to reach a minimum value of 68.64 $W.m^{-2}.K^{-1}$ after 300 min of heating.

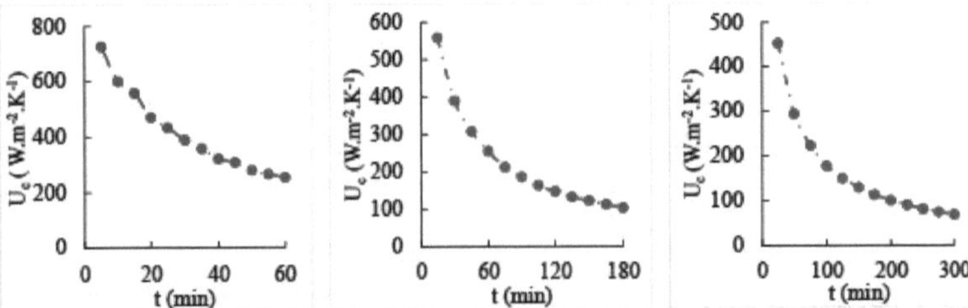

Figure 4. 5: Variation in heat transfer coefficient as a function of heating time

2.1.4 Temperature field

Figure 4.6 shows the temperature distribution of the water in the tank for the three different heating times defined by t=60 min, t=180 min and t =300 min. From the results obtained, it can be mentioned that temperature stratification in the lower part is formed. After 60 min of heating, three zones are observed; the first and second zones are located in the lower part of the tank, while the third zone is formed near the condenser. It is located in the middle and upper part of the tank. As the heating time increases, the number of zones increases and is formed due to the difference in density. A comparison of these results confirms that the minimum and maximum water temperatures increase linearly when we continue to make use of the heat rejected by the condenser coil. According to the simulation results, the difference in water temperature between the lower and upper parts of the tank increases as the heating time increases. During the heating process, hot water continuously flows upwards and accumulates in the upper part of the tank to form a hot zone. This is because the water in the upper part of the tank is gradually heated by natural convection from the refrigerator's condenser. This confirms that the temperature of the water in the upper part of the water tank is obviously higher than in the lower part.

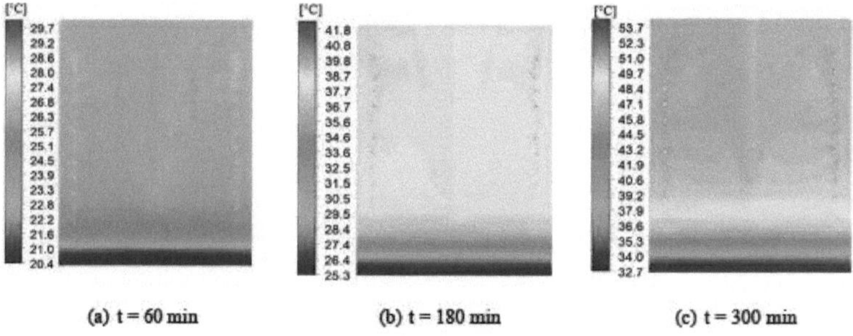

Figure 4. 6: Temperature distribution

2.1.5 Speed field

Figure 4.7 shows the variation in water velocity over three heating periods defined by t=60 min, t=180 min and t=300 min. For all three heating periods, it can be seen that the water flows at high velocity in the middle and upper part of the tank, indicating the amount of heat transferred by convection in these two zones. If we look at the distribution of the water velocity field in the tank, we can see that the velocity of the water in the middle and upper part of the tank is obviously higher than in the lower part. Under these conditions, the maximum value is found in the centre line of the tank, which is close to the condenser coil. With the increase in heating time at t=60 min and t=180 min, the velocity distribution is similar with the exception of the profile in the middle part of the tank which is due to the importance of the heat transfer phenomenon generated by the tube wall. After 300 min of heating, the velocity field shows a small variation compared with the other heating times.

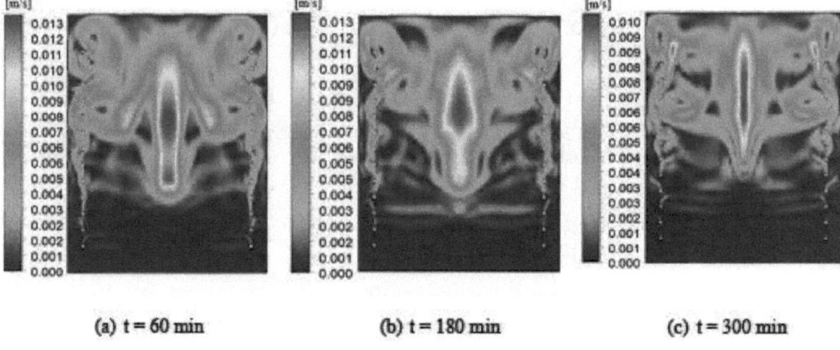

Figure 4. 7: Speed distribution

2.2 Parametric study

A parametric study of the helicoidal condenser is developed in this section. This involves varying one parameter while setting the others. The aim is to study how this variation affects the thermal performance of a domestic refrigerator coupled with a water heating unit, by exploiting the waste heat given off by the condenser.

2.2.1 Effect of condenser pitch

2.2.1.1 Speed profile

The variation in water velocity as a function of heating time is shown in Figure 4.8. From these results, it can be seen that the distance between the condenser steps is an important factor influencing the thermal performance of a refrigeration machine used to produce domestic hot water. The simulation results obtained show that the water velocity has a maximum value when using a condenser with a pitch equal to h=20 mm.

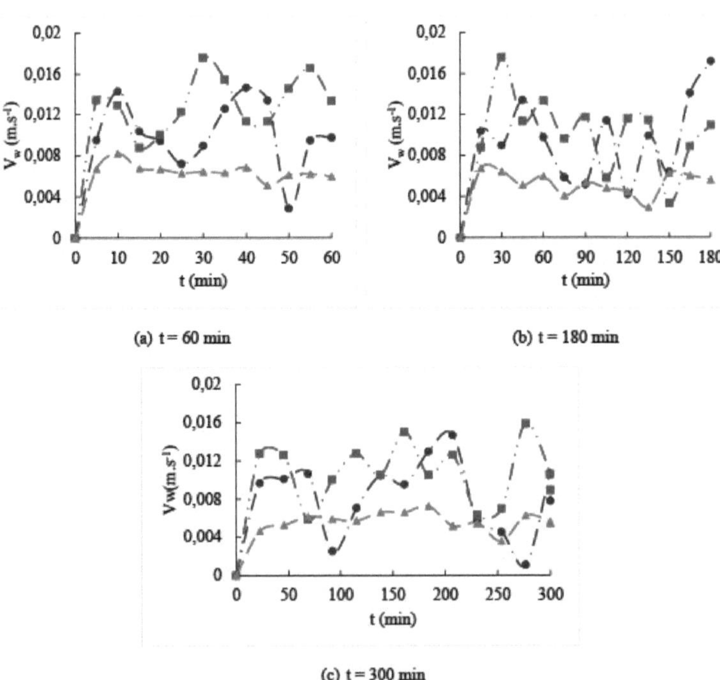

Figure 4. 8: Variation of water velocity as a function of time for three different steps

2.2.1.2 Speed field

From Figure 4.9, the helicoï'dal condenser with a pitch equal to h=20 mm shows better thermal performance than the other two types of condenser, namely h=10 mm and h=30 mm. It can also be seen that the high speed on the centre line of the tank is due to the better heat transfer coefficient provided by the condenser with a pitch defined by h=20 mm. It is therefore clear that the pitch of the helicoidal condenser is an important parameter influencing the operation of a domestic refrigerator coupled with a hot water production unit.

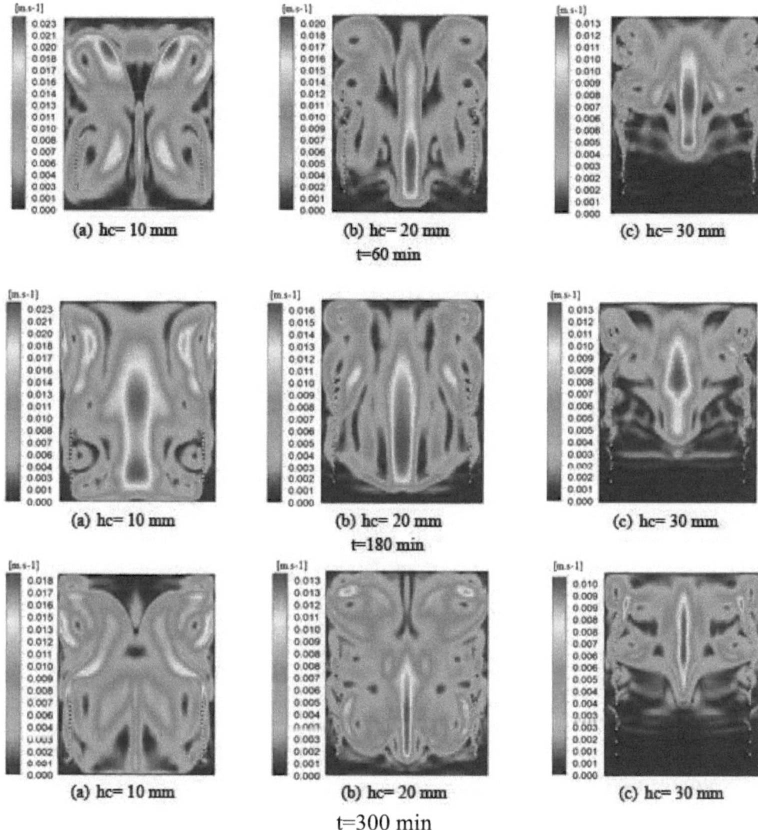

Figure 4. 9. speed contours

2.2.1.3 Temperature profile

The influence of the pitch of the helicoidal condenser on the variation of the water temperature inside the tank is shown in Figure 4.10. The results of the numerical simulation show that the water temperature increases as the condenser pitch is reduced for the same heating time. From these simulation results, it can be seen that the low-pitch condenser causes an increase in convective heat transfer. This can be explained by the relationship between the heat transfer coefficient and the condenser geometry. To examine the influence of this parameter on thermal performance, the results obtained show that the water temperature has a maximum value when using a condenser with a pitch equal to h=20 mm.

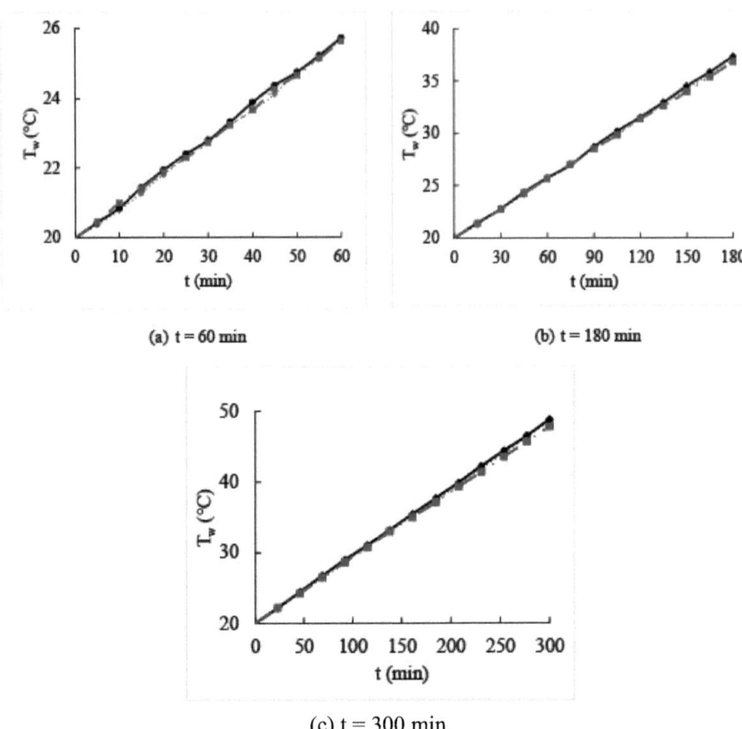

(c) t = 300 min

Figure 4. 10. Temperature variation as a function of condenser pitch

2.2.1.4 Temperature field

The variation in water temperature as a function of three different condenser pitches is shown in Figure 4.11. These results indicate that condenser geometry is a main parameter affecting the water temperature inside the tank. For the three heating periods, the highest temperature occurs for a condenser with a pitch equal to h=20 mm.

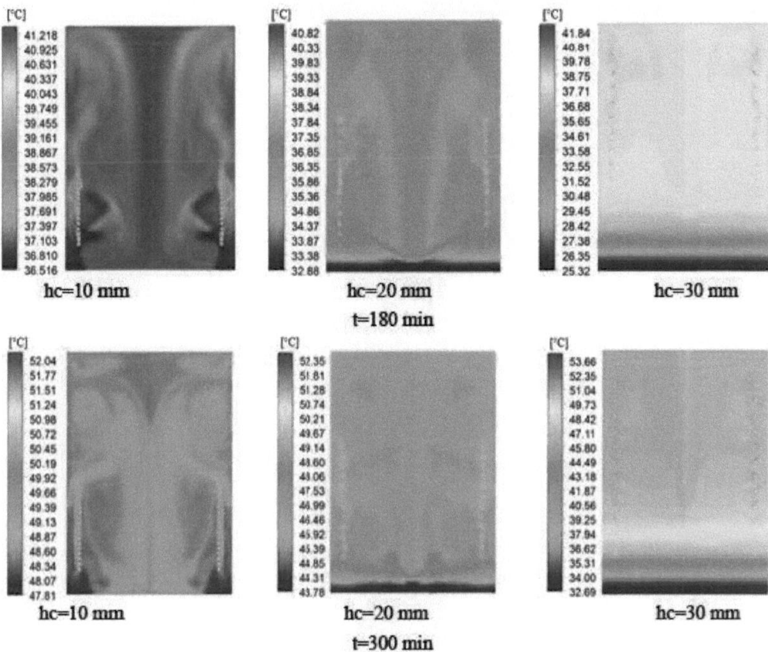

Figure 4. 11. Temperature contours

2.2.1.5 Pressure field

Figure 4.12 shows the distribution of the pressure field inside the water tank during three heating periods defined by t=60 min, t=180 min and t=300 min. This study makes it possible analyse and compare the impact of condenser geometry on the thermal performance of a domestic refrigerator coupled with a domestic water heating unit. The results show that increasing the pitch of the condenser reduces the heat transfer coefficient between the tube wall and the water, thereby increasing heat loss. Furthermore, it can be seen that the condenser with a hc=20 mm pitch provides the highest heat flow compared with the other geometries during the same operating time, given that the heat transfer zone is larger.

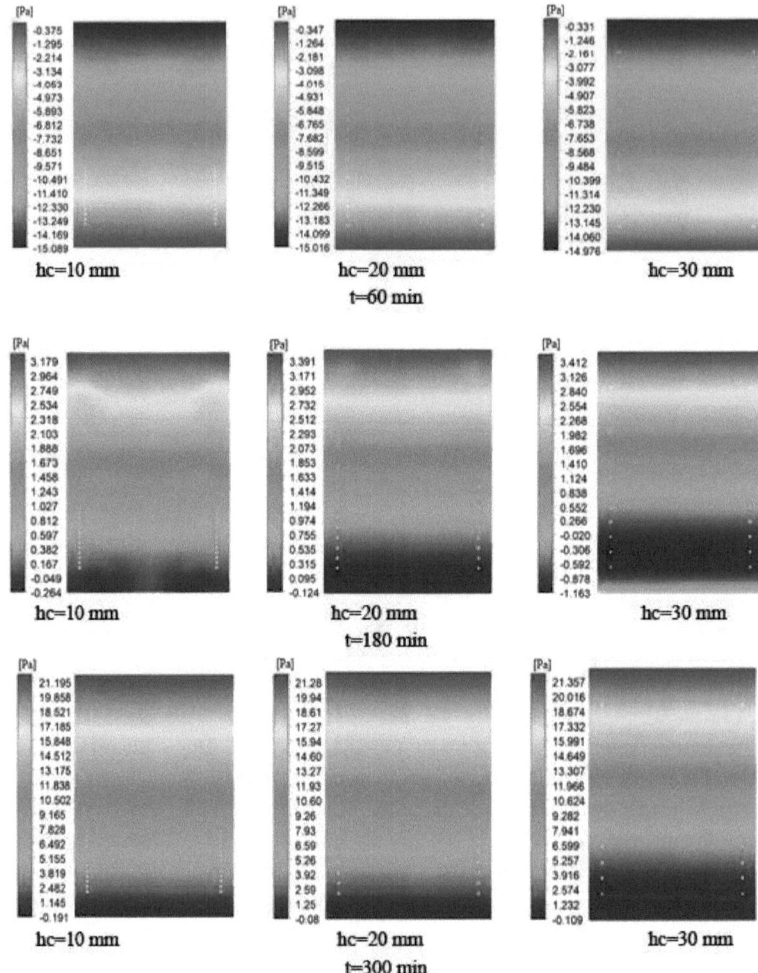

Figure 4. 12. Pressure contours

2.2.2 Effect of condenser tube diameter

2.2.2.1 Speed profile

Figure 4.13 shows the effect of condenser tube diameter on the thermal performance of a domestic refrigerator coupled with a water heater. During the three heating periods, the results obtained show that as the tube diameter increases, the water velocity also increases. In addition, it can be seen that the tube diameter of dc = 6 mm ensures a lower velocity than the other diameters for the same time. These results indicate that the diameter of the condenser tube is the main parameter affecting water velocity.

dc= 6 mm   dc= 8 mm   dc= 10 mm

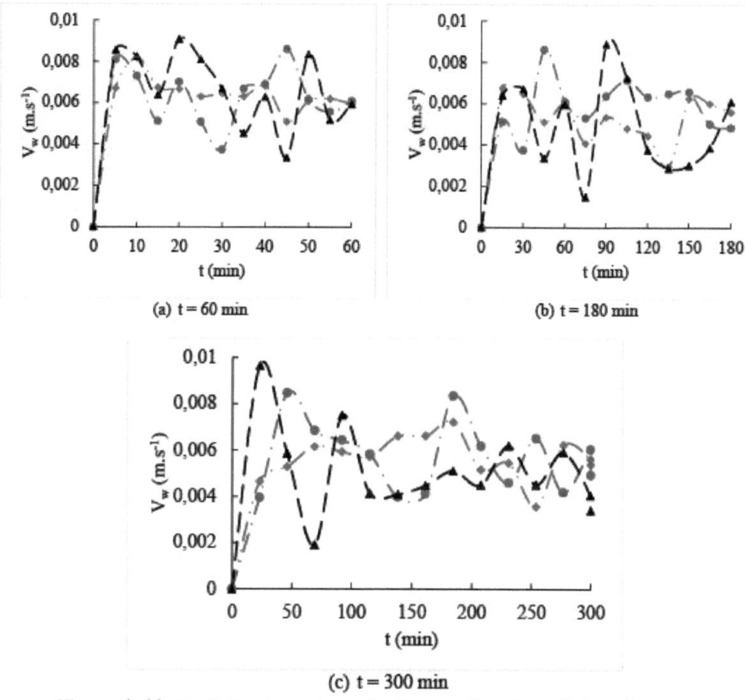

Figure 4. 13: Variation in water velocity as a function of pipe diameter

2.2.2.2 Speed field

Figure 4.14 shows the effect of the condenser tube diameter on the velocity field distribution during the three heating periods. The results of the numerical simulation show that increasing the tube diameter leads to an increase in the velocity for the same heating period. By increasing the diameter of the tube from dc= 6mm to dc=10 mm, there is a progressive increase in the velocity within the upper part of the tank. These results confirm that the tube diameter is a main factor affecting the thermal performance of a domestic refrigerator coupled with a water heater.

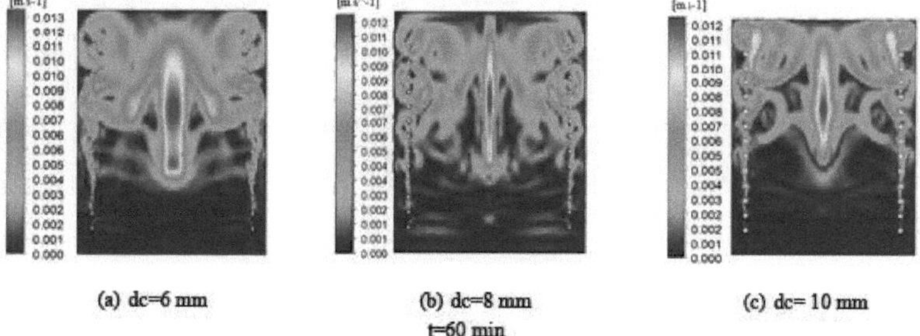

(a) dc=6 mm   (b) dc=8 mm   (c) dc= 10 mm

t=60 min

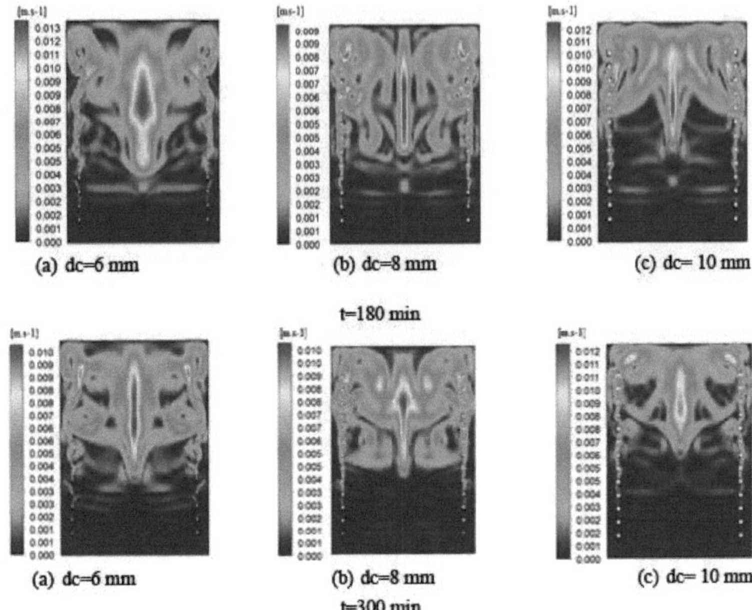

Figure 4. 14. Speed distribution

2.2.2.3 Temperature profile

Figure 4.15 shows the variation in water temperature as a function of the diameter of the condenser tube. From these results, it can be seen that increasing the diameter of the tube increases the temperature of the water inside the tank. It can also be seen that the temperature of the water in the tank is highest when the tube with dc=10 mm is used. Under these conditions, the water is heated to 26°C in 38 min.

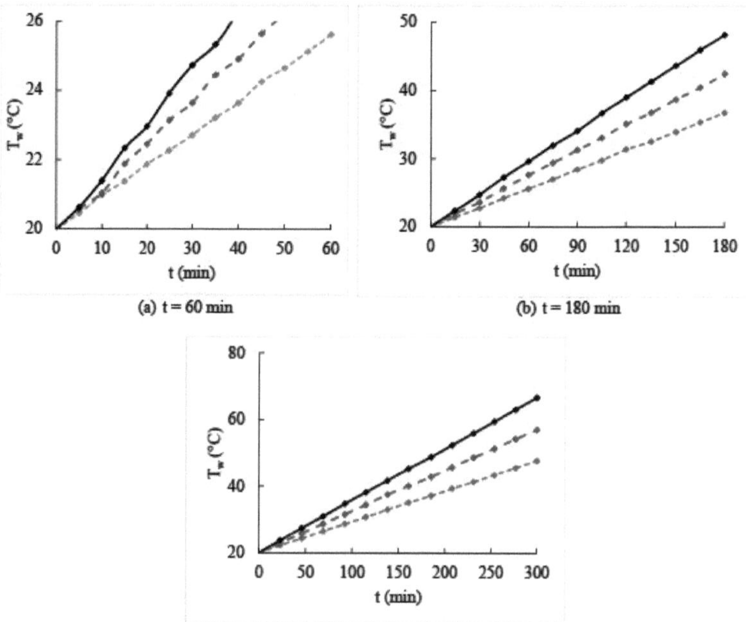

Figure 4. 15. Variation in water temperature as a function of condenser tube diameter.
condenser

2.2.2.4 Temperature field

Figure 4.16 shows the water temperature profile for different condenser tube diameters. From these results, we can see that the water temperature varies according to the tube diameter. At t=60 min, it should be noted that for a tube diameter dc=10 mm, the maximum temperature is reached, i.e. 30°C. As the diameter of the condenser increases, the temperature of the water in the upper part of the tank rises and the stratification phenomenon begins to diminish. In addition, these results indicate that the temperature gradually decreases from dc =10 mm to dc = 8 mm.

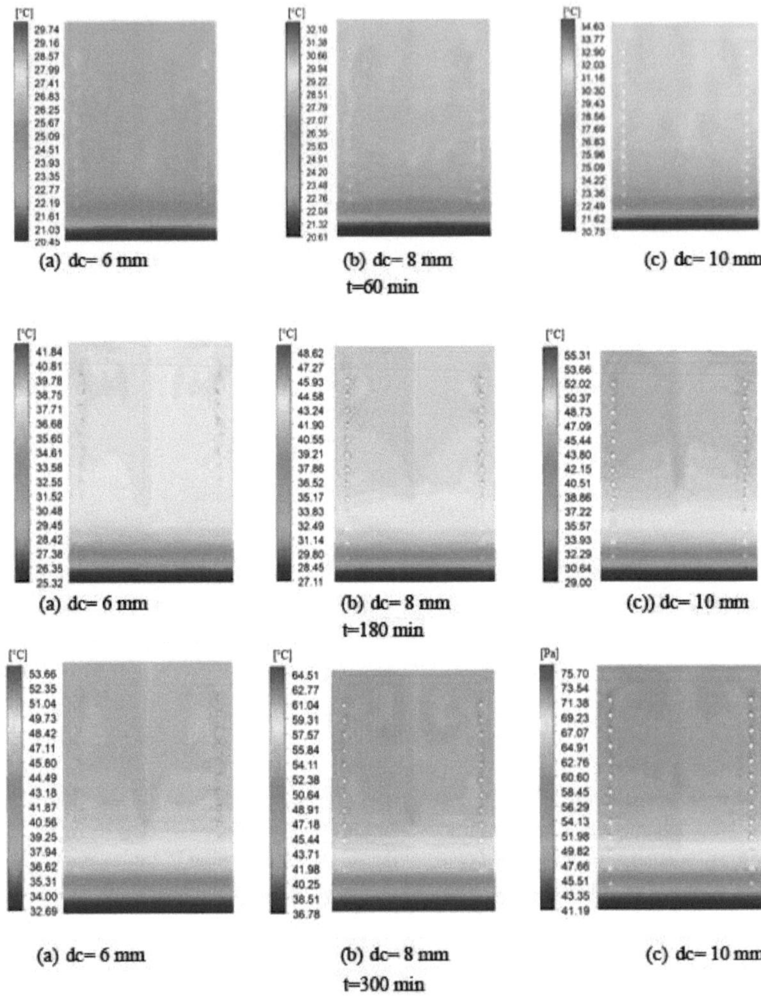

Figure 4. 16. Temperature distribution

2.2.2.5 Pressure field

The distribution of the pressure field in the tank is shown in Figure 4.17. From these results, it can be seen that the pressure increases progressively with increasing tube diameter over a period equal to t=60 min, t=180 min and t=300 min. A clear improvement in the pressure field evaluated at 41.22% is observed in these results. In addition, a moderate decrease in pressure was observed for the small diameter value equal to dc=6 mm. In fact, the impact of the tube diameter is visible for dc=6 mm. Under these conditions, a better pressure distribution is obtained compared to the diameter dc =10 mm.

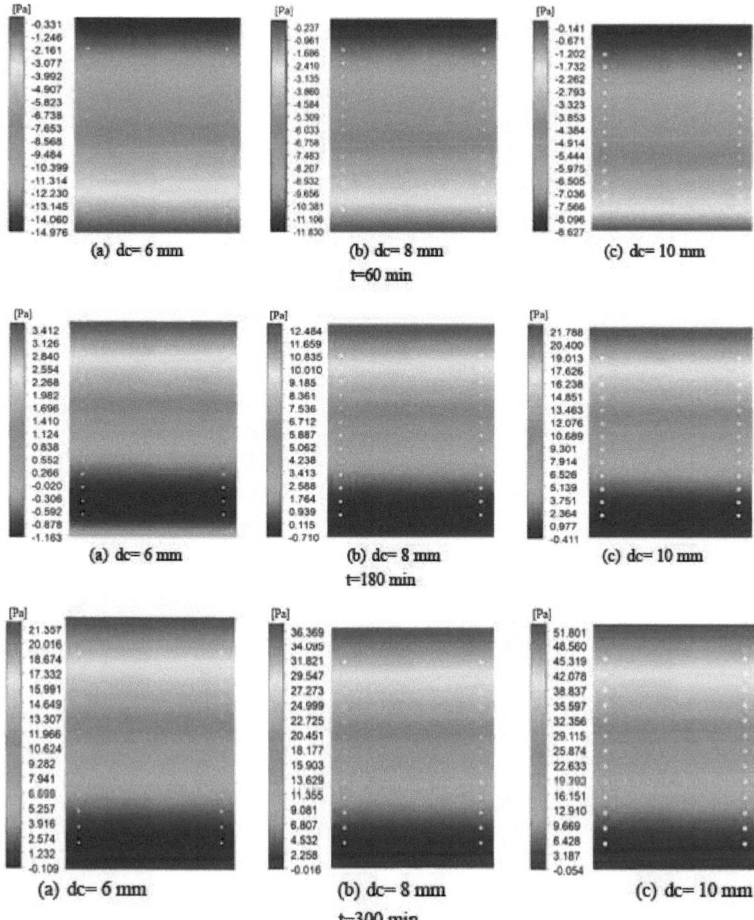

Figure 4. 17. Pressure contours

2.2.3 Effect of the number of condenser turns

2.2.3.1 Speed profile

The influence of the number of helicoidal condenser turns on the variation in water velocity is shown in Figure 4.18. In this study, a comparison was made of the velocity profiles for three different numbers of turns represented by N=11, N=13 and N=15. The results obtained show that the velocity increases as the number of condenser turns decreases for the same heating period. This increase can be explained by the increase in the amount of thermal energy produced by the condenser in the lower part of the water tank. From the results of the numerical simulation, it can be seen that increasing the diameter of the condenser and reducing the number of turns results in significant heat exchange between the water and the condenser wall with minimal heat loss.

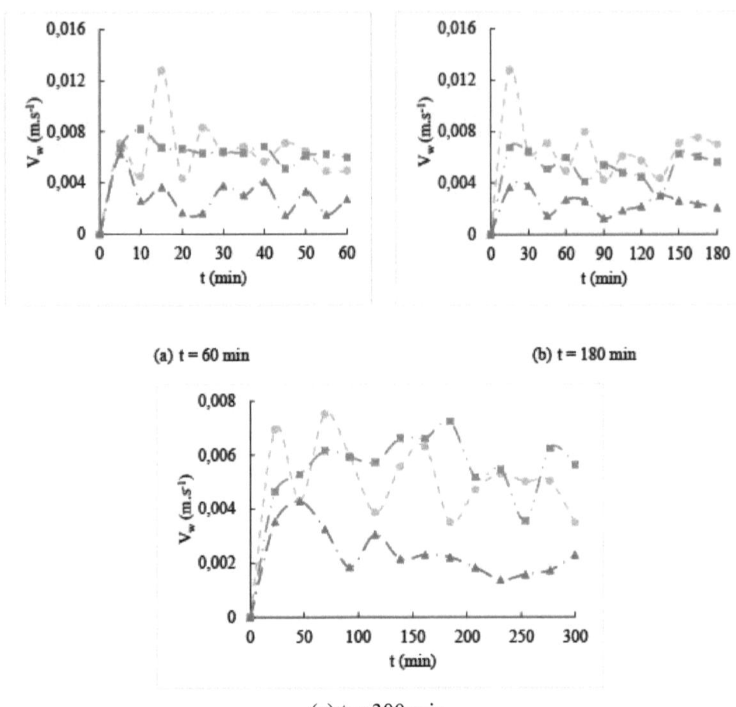

(c) t = 300 min

Figure 4. 18. Variation in water velocity as a function of the number of condenser turns

2.2.3.2 Speed field

A comparison of the distribution of the water velocity field for different numbers of turns of the helicoidal condenser is shown in Figure 4.19. From these results, it can be seen that for t= 60 min, the water velocity decreases with the variation in the number of turns from N= 11 to N= 15. It can also be seen that heat exchange along the condenser varies from one zone to another, as indicated by these results. This is mainly due to the reduction in the number of condenser turns and the increase in the total diameter. The decrease in the number of turns and the increase in the diameter of the condenser mean that there is good turbulence in the upper part of the tank. This results in good heat transfer by natural convection. Finally, we note that the highest water velocity corresponds to a helical condenser with N=11.

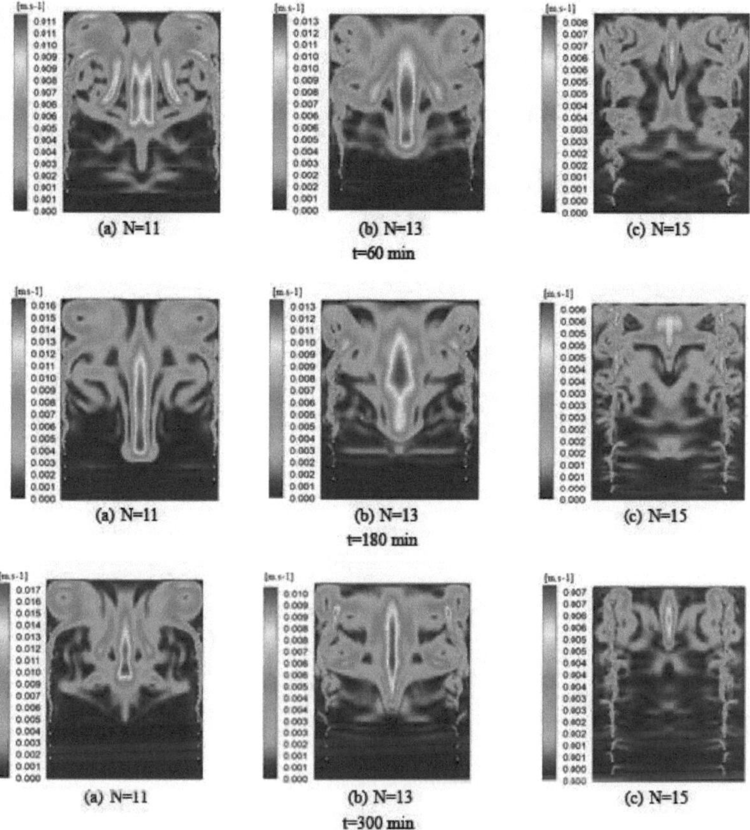

Figure 4. 19. Speed contours

2.2.3.3 Temperature profile

Figure 4.20 shows the variation in water temperature as a function of the number of turns in the helicoidal condenser. These temperature profiles obtained during the heating process also show the impact of changing the number of turns for the three configurations considered, N=11, N=13 and N=15. To analyse the effect of condenser geometry on thermal performance, a comparison is made between the numerical results obtained for t=60 min, t=180 min and t=300 min. It should also be noted that the average temperature of the water increases considerably when the number of turns of the helicoidal condenser is increased. This can explained by the increase in temperature due to the increase in the vertical surface area of the condenser in the water . These same results were described earlier in study of speed variation.

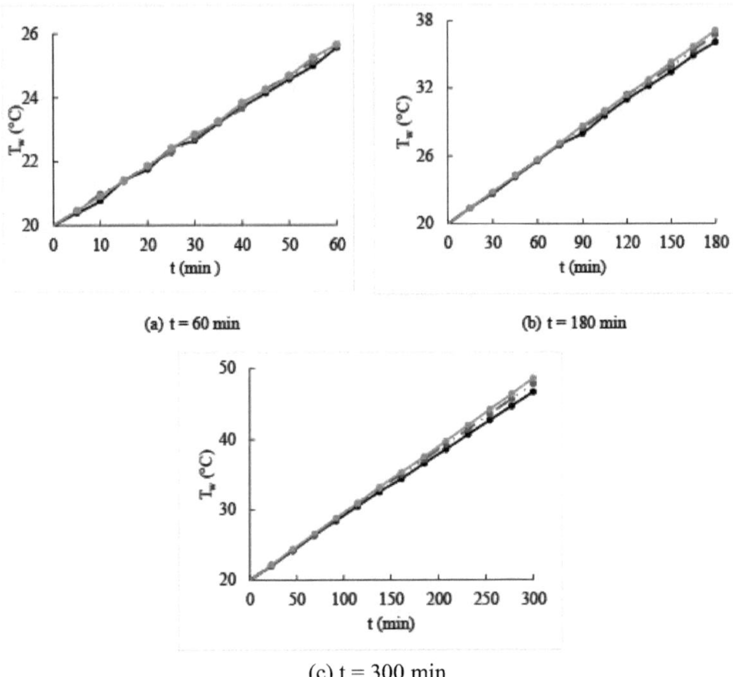

(c) t = 300 min

Figure 4. 20. Variation in water temperature as a function of condenser tube diameter. condenser

2.2.3.4 Temperature field

Figure 4.21 shows the temperature distribution of the water in the receiver for the three configurations N=11, N=13 and N=15 turns. The results obtained show that the reduction in the number of turns is very effective because the flow of refrigerant inside the condenser tube increases the speed of the water where the temperatures

high. These results also show the formation of recirculation zones in the upper part of the tank, making it possible to quantify the convective exchanges between the walls of the condenser and the water circulating in it.

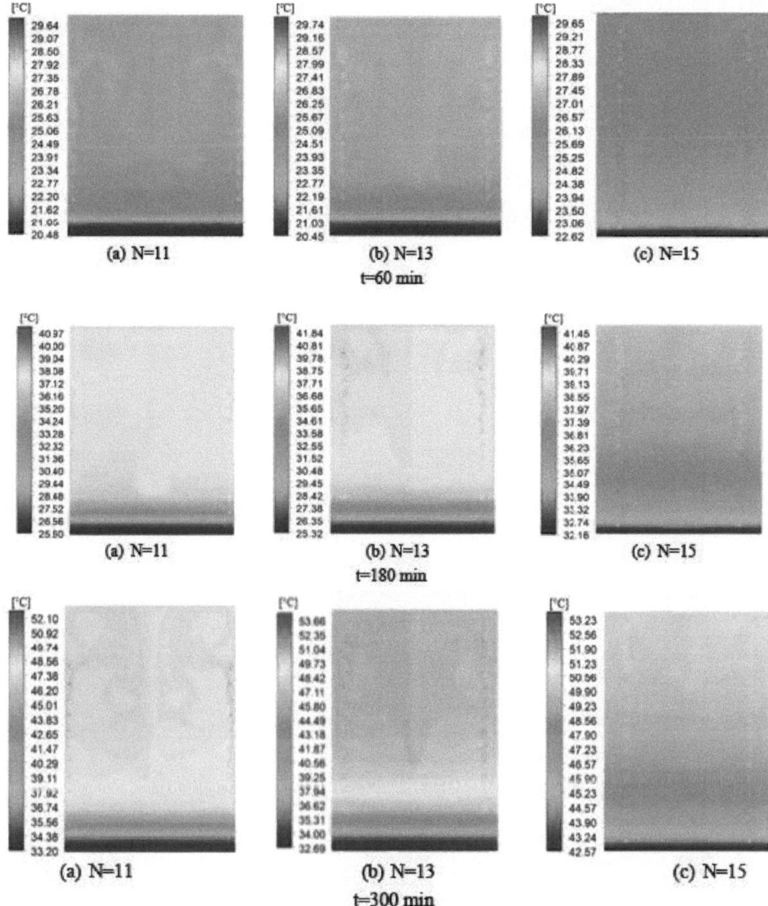

Figure 4. 21.Temperature contours

2.2.3.5 Pressure field

Figure 4.22 shows the distribution of the pressure field in the tank for the three configurations N=11, N=13 and N=15 turns. From the results of this study, it can be seen that the pressure decreases as the number of turns of the helical condenser increases from N=11 to N=15. The negative sign of the pressure values corresponds to the change in the direction of water flow in the tank, which explains the presence of a negative pressure zone. These results also show that the use a helical condenser with N=11 turns is therefore beneficial for improving heat exchange between the condenser walls and the water.

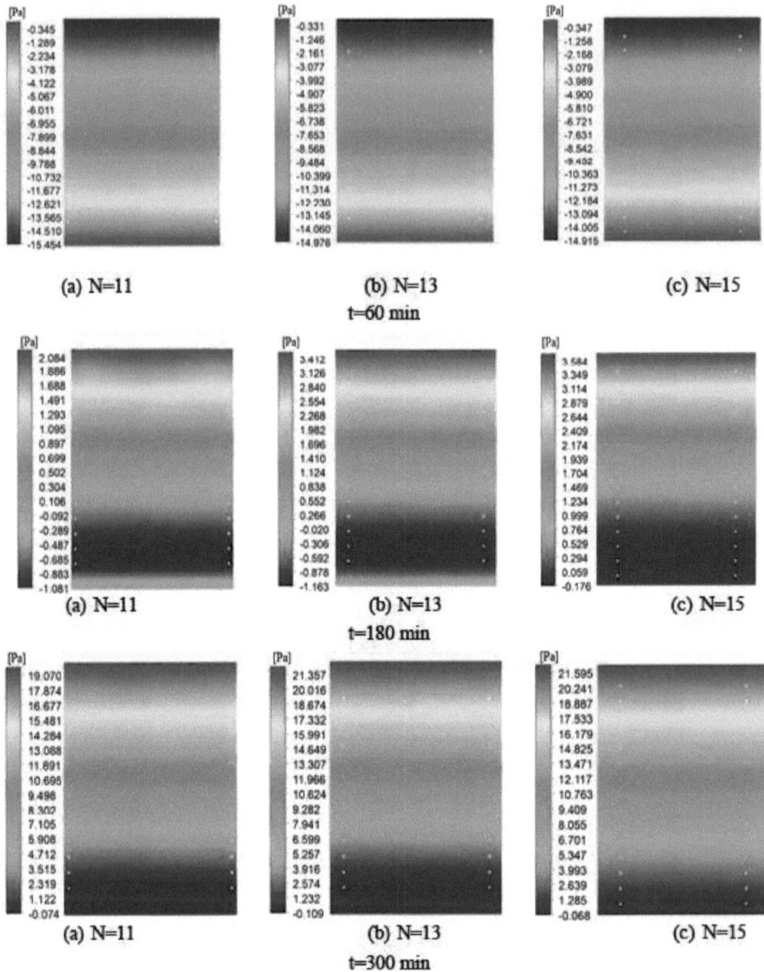

(a) N=11 (b) N=13 (c) N=15

t=60 min

(a) N=11 (b) N=13 (c) N=15

t=180 min

(a) N=11 (b) N=13 (c) N=15

t=300 min

Figure 4. 22. Pressure contours

2.3 Numerical optimisation of the helical condenser geometry

The helicoidal condenser is widely used in refrigeration machines for the production of domestic hot water because of its energy efficiency, simplicity of manufacture and improved heat transfer. It is therefore of interest to develop and optimise helicoi'dal condenser tube designs to system performance, heat transfer and temperature stratification in the tank. Therefore, the objective of this study is to improve the thermal performance of a helical condenser by modifying the tube shape. Thus, this section focuses on the influence of the helical condenser tube shape on thermal performances such as velocity, temperature and heat transfer coefficient. In addition, obtaining the optimum structure of the condenser tube and improving the heat transfer process between the condenser tube and the water is of interest for the operation of the domestic refrigerator with immersed condenser for the production of hot water.

2.3.1 Comparison between variable and constant pitch

2.3.1.1 Speed profile

The influence of the condenser geometry on the water velocity inside the tank is shown in Figure 4.23. These numerical results show the evolution of the water velocity when using a helicoidal condenser with a constant pitch and another with a variable pitch. The effect of this parameter was studied for three different times defined by t=60 min, t=180 min and t=300 min. To analyse and compare the impact of this parameter on the velocity distribution, the total height, length and diameter of the two types of condenser are equal. Also, the geometric parameters of the water storage tank are kept constant. After 1 h of heating, it can be seen that the value of the speed has increased with the use of a variable-pitch helical condenser. Under these conditions, the maximum velocity reaches $V=0.017\ m.s^{-1}$. By following the heating process up to t= 3h, the speed of the water increased further and the difference in maximum speed also increased. Furthermore, it can be seen that the use of a variable-pitch helical condenser causes an increase in the heat transfer coefficient inside the tank, which in turn leads to an increase in water velocity. By increasing the heating time to 300 min, the difference is obviously significantly increased. Consequently, there are clear effects of condenser geometry on water velocity. For a maximum heating period, t=300 min, the fields of water velocity versus tank height indicate that with use of the variable pitch helicoidal condenser, the heat transfer area in the lower and middle parts of the tank is increased. This fact can increase the flow of heat to the lower part of the tank and improve thermal performance. This confirms that natural convection becomes important with the use of a variable pitch helical condenser. In fact, when using a constant-pitch condenser for water heating, the lower part of the water does not favour natural convection with the upper part and the speed of the water inside becomes very low. This leads to thermal stratification, which can be avoided with a variable-pitch condenser. Comparison of these results confirms that the shape of the coil in the helicoidal condenser has a direct effect on the distribution of water velocity.

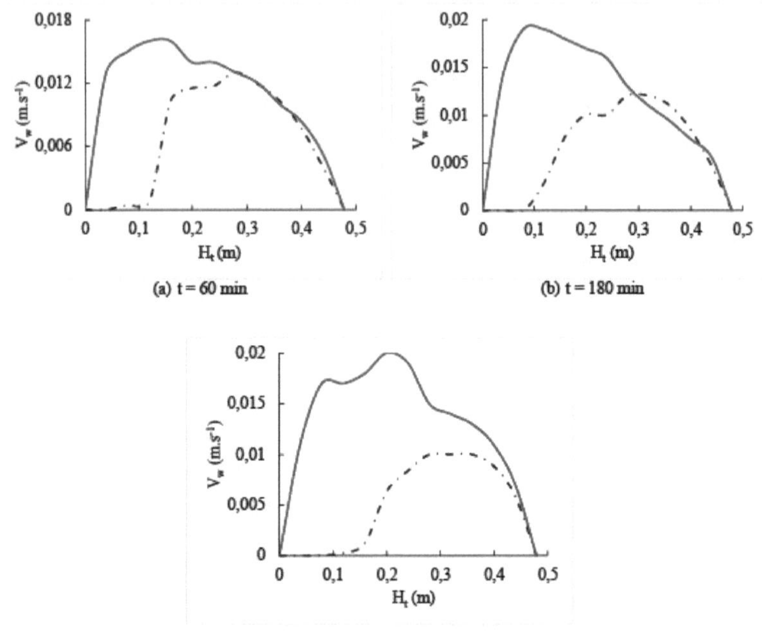

(c) t = 300 min

Figure 4. 23.Variation in water velocity as a function of condenser geometry **2.3.1.2 Temperature profile**

The evolution of water temperature as a function of tank height for t=60 min, t=180 min and t=300 min is shown in Figure 4.24. For the same operating condition, the temperature profile was plotted for two types of condenser, namely the constant pitch helical condenser and the variable pitch helical condenser. This study shows the impact of condenser geometry on the variation in water temperature inside the tank. The results obtained from the study indicate that the geometry of the condenser is a main parameter affecting the temperature of the water during the heating process. Analysis of these results shows that the water temperature increases with the height of the tank. After 1 hour of heating, the water temperature in the upper and middle part of the tank is uniform and reaches the value T=26.29 °C. However, the temperature gradually decreases from T=26.29 °C to T=20.46 °C from the top of the tank to the bottom, resulting in a maximum temperature difference of 5.83 °C and 0.04 °C, respectively, for the constant-pitch and variable-pitch condenser. Thus, it can be deduced that with the use of a variable pitch condenser, the difference in water temperature on the tank centre line is reduced. In addition, these results indicate that, due to the extent of heat transfer by the variable-pitch condenser, the water temperature has spread evenly from the top to the bottom of the tank. As a result, the water temperature rises uniformly when the variable-pitch condenser is used. This can explained by the fact that the heat transfer provided by the variable-pitch condenser is relatively high in the lower part of the condenser. As the heating time increases, the water temperature also increases. For t=180 min, the average water temperature is increased and the maximum temperature difference on the tank centre line is 13.25 °C and 2.72 °C for the constant-pitch condenser and the variable-pitch condenser respectively. At t=300 min, the effects of condenser geometry on thermal behaviour are still evident. Compared with the constant-pitch condenser configuration, the variable-pitch condenser has the highest temperature profiles. In addition, the temperature of the water in the tank is almost uniform and thermal stratification is virtually eliminated thanks to the new design of this variable-pitch condenser.

Condenseur cylindrique — Condenseur cylindrique à pas variable

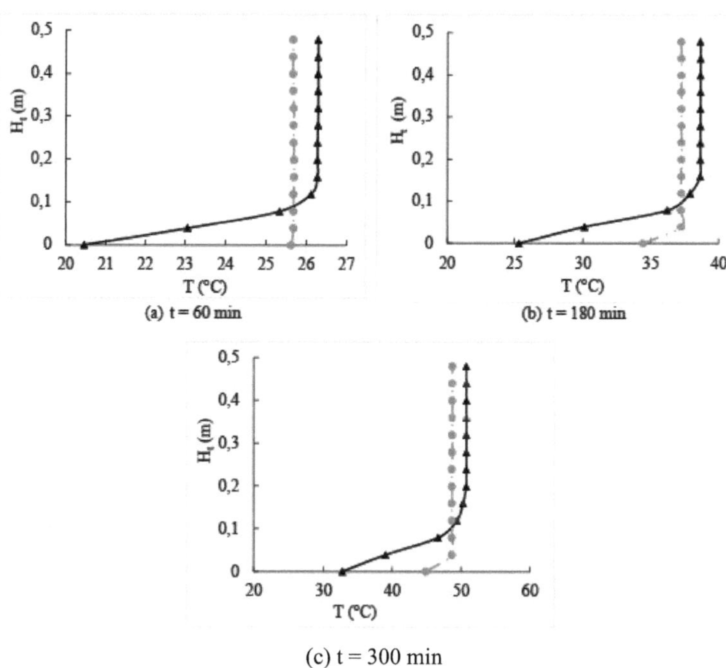

(c) t = 300 min

Figure 4. 24. Variation in water temperature as a function of condenser geometry

2.3.1.3 Heat transfer coefficient profile

In order to qualitatively analyse the two proposed configurations of the helicoidal condenser, the profiles of the heat transfer coefficients at three different heating times defined by t=60 min, t=180 min and t=300 min are presented in Figure 4.25. From these results, it should be noted that the two condenser configurations show similar results with a decrease in the heat transfer coefficient with increasing heating time. At t=60 min, the average heat transfer coefficient for the variable-pitch condenser is 424.60 W/m^2.K, whereas for the constant-pitch condenser it is 413.47 W/m^2.K. As the heating time increases from t=60 min to t=300 min, the average heat transfer decreases and the maximum difference between the heat coefficients of the two configurations increases. In fact, when the variable pitch helical condenser is used, the heat transfer increases. This is mainly due to its relatively high heat flow in the lower part resulting from the small distance between the condenser pitches. It can therefore be deduced that the variable pitch condenser has the highest heat transfer coefficient compared to a constant pitch helical condenser.

Condenseur cylindrique Condenseur cylindrique à pas variable

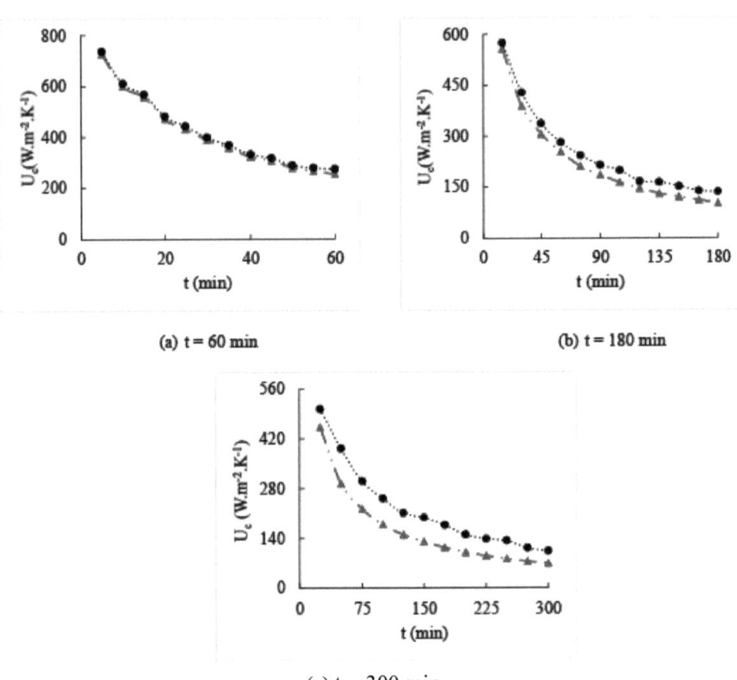

(c) t = 300 min

Figure 4. 25. Variation of heat transfer coefficient as a function of condenser geometry condenser

2.3.1.4 Speed field

Figure 4.26 shows the variation in water velocity as a function of heating time for the two condenser configurations with constant and variable pitch. To examine the influence of this geometrical parameter on the variation in velocity, three different heating periods were considered, defined respectively by t=60 min, t=180 min and t=300 min. Examination of these results shows that the speed is very low for the constant-pitch condenser compared with the second variable-pitch geometry. After one hour's operation, the maximum water velocity is 0.013 m/s and 0.016 m/s for the constant-pitch condenser and the variable-pitch condenser respectively. With the increase in the heating time from t=60 min to t=180 min, the maximum speed is 0.020 m/s for the variable-pitch condenser, whereas it remains the same for the constant-pitch condenser. In fact, it should be noted that convection is very important with the use of a variable-pitch helical condenser. This fact confirms the choice of the helical condenser to ensure the best performance. At t=300 min, the maximum speed is equal to

0.010 m/s and 0.020 m/s for the constant-pitch and variable-pitch condenser respectively. In addition, these results indicate that the variable-pitch condenser improves the convective transfer coefficient between the water in the receiver and the refrigerant in the condenser. This is mainly due to the reduction in the condenser pitch in the lower part of the tank. Furthermore, the results obtained may lead us to improve the natural convection of the lower part of the condenser. Thus, the comparison between these results confirms that the geometry of the condenser has a direct effect on the distribution of the water velocity field, which in turn contributes to improving the temperature in the storage tank.

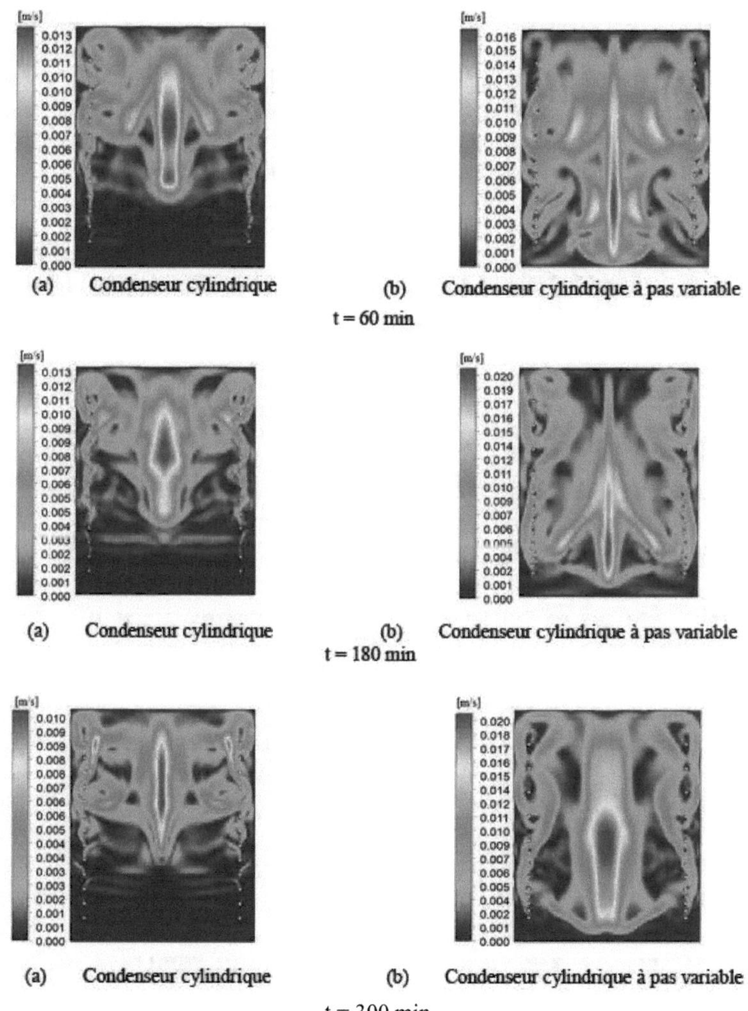

Figure 4. 26. Speed distribution

2.3.1.5 Temperature field

In this section, the water temperature distribution is presented for three heating times defined by t=60 min, t=180 min and t=300 min. The results show the impact of the condenser geometry on the temperature distribution. After 60 min of heating, the average water temperatures for a constant-pitch condenser and a variable-pitch condenser were T=25.08 °C and T=26.61 °C respectively. In this study, the water temperature of these two configurations is compared. The calculations show that the phenomenon of thermal stratification is remarkable when a constant-pitch condenser is used. For the same heating period, thermal stratification decreases with the use of a variable-pitch helical condenser. It is also interesting to note that the temperature of the water inside the tank

affected by the geometry of the condenser. With the increase in heating time from t=60 min to t=180 min, it can be seen that the water temperature distribution is affected by the condenser geometry. In fact, the water temperatures in the lower part of the tank for the constant pitch and variable pitch condenser are equal to T=25.3°C and T=34.4°C respectively. Because of the difference in temperature in the lower part of the tank, thermal stratification cannot be improved by using a constant-pitch condenser. The results show that the degree of stratification depends mainly on the geometry of the condenser and the heating time. Under these conditions, it can be seen that the maximum difference between the water temperature distribution for a constant-pitch and variable-pitch condenser increases with increasing heating time from t=60 min to t=300 min. As a result, the water temperature in the tank is uniform and the phenomenon of thermal stratification is completely eliminated thanks to the new geometry used.

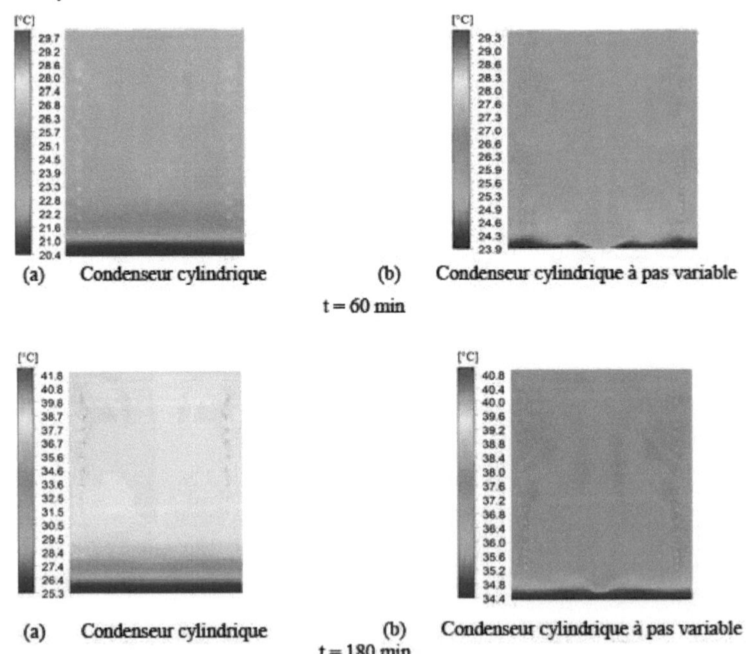

(a) Condenseur cylindrique (b) Condenseur cylindrique à pas variable

t = 60 min

(a) Condenseur cylindrique (b) Condenseur cylindrique à pas variable

t = 180 min

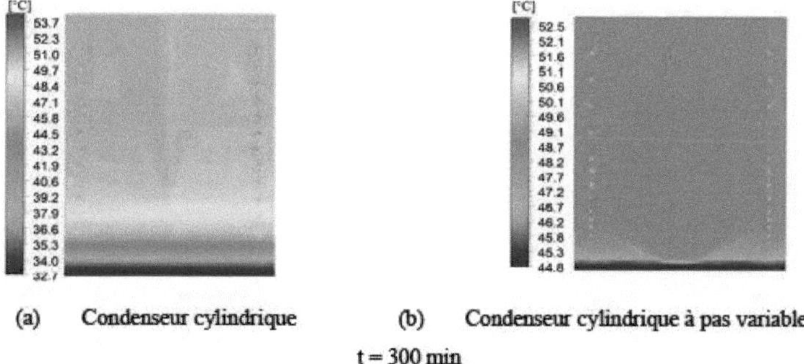

Figure 4. 27. Temperature contours

2.3.1.6 Pressure field

It is clear that the geometry of the condenser is an important parameter influencing the operation of a domestic refrigerator coupled with a water heater. Consequently, it is interesting to study the effect of this parameter on the distribution of the pressure field inside the water tank. Figure 4.28 shows that the pressure increases advantageously with the use of a variable pitch helical condenser. As these results indicate, a variable-pitch helical condenser thus shows better thermal performance than a constant-pitch condenser.

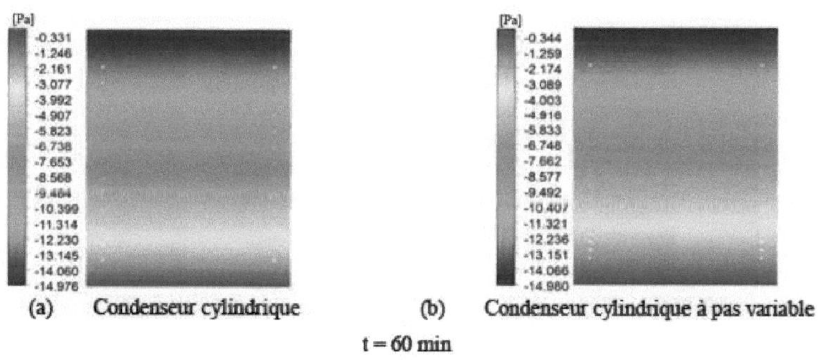

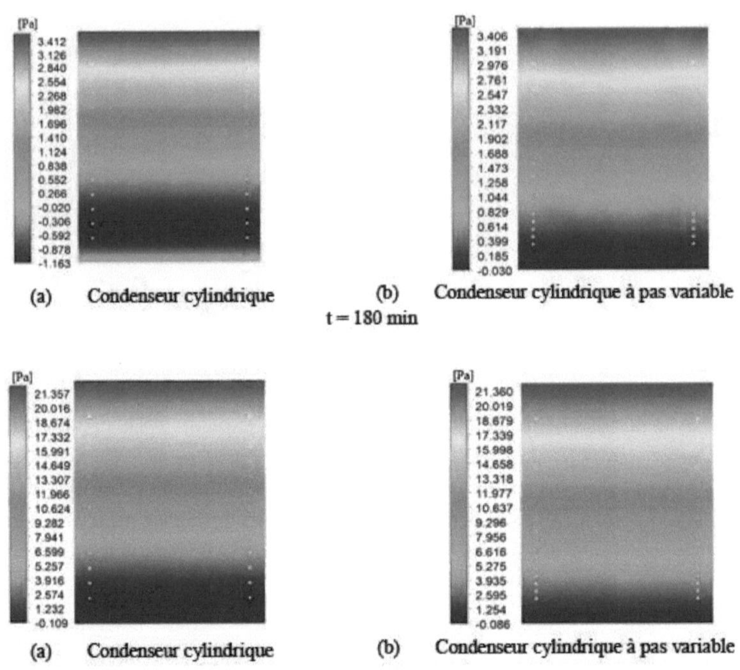

t = 300 min

Figure 4. 28. Pressure contours

2.3.2 Effect of diameter

This section compares the performance of a variable-diameter helicoidal condenser with a conical shape and a constant-diameter condenser.

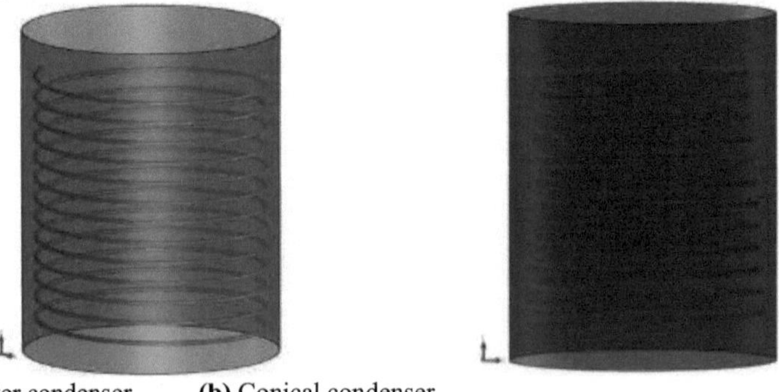

(a) Diameter condenser constant **(b)** Conical condenser

Figure 4. 29. The geometry of the constant and variable diameter condensers

2.3.2.1 Speed profile

The influence of the condenser geometry on the variation in water velocity is shown in Figure 4.30. The results show that the external shape of the helicoidal condenser totally immersed in the water tank affects the variation in velocity. The calculations confirm that the use of a cone-shaped condenser to heat the water is therefore beneficial for improving the heat transfer phenomenon between the tube wall and the water on its outside. With the increase in heating time from t=60 min to t=300min, the influence of geometry becomes clearer.

- · - Condenseur cylindrique — · · Condenseur conique

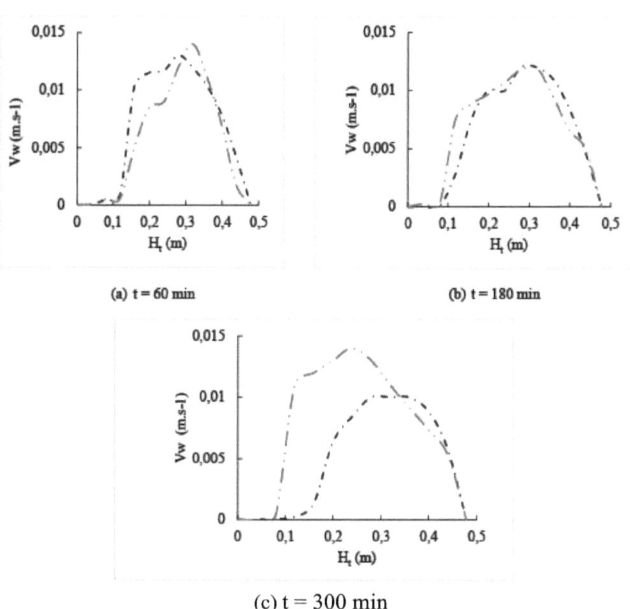

(c) t = 300 min

Figure 4. 30. Variation in water velocity as a function of condenser geometry

2.3.2.2 Temperature profile

Figure 4.31 shows the variation of water temperature with tank height for the cylindrical and conical shaped condensers for three heating times defined by t=60 min, t=180 min and t=300 min. At t=60 min, the average water temperatures for the cylindrical condenser and the conical condenser are T=25.50 °C and T=25.53 °C respectively. This difference in temperature is mainly due to the increase in the overall diameter of the condenser at the bottom of the tank. As the heating time increases, the average temperature of the water increases and so does the difference between the two results. The results confirm that the conical condenser shape improves the heating process and provides a higher heat flow than the cylindrical shape. Recovering the heat given off by a conical condenser increases the temperature of the water to 50.73°C. As a result, this hot water can be used to meet domestic and industrial needs such as cleaning and washing.

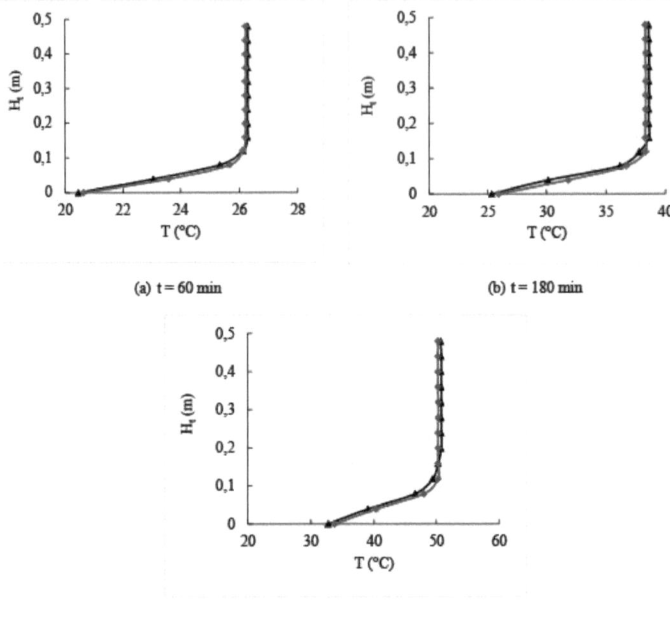

(c) t = 300 min

Figure 4. 31. Variation in water temperature as a function of condenser geometry

2.3.2.3 Heat transfer coefficient profile

Figure 4.32 shows the variation of the heat transfer coefficient as a function of time for a cylindrical and a conical condenser. From the results obtained, it can be seen that increasing the heating time from t=60 min to t=300 min causes an increase in the difference in the heat transfer coefficient between the helical and conical shaped condenser. Furthermore, increasing the heating time reduces the heat transfer coefficient between the condenser wall and the water, as shown by the results obtained. At the start of operation, the heat transfer coefficient is equal to 700.38W/m^{-2}.K^{-1}. It then decreases progressively until it reaches a value of 70.43 W/m^{-2}.K^{-1} after a period of t=300min. This can be explained by the decrease in heat transfer between the condenser and the water due to the increase in the temperature of the water inside the tank. It can also be seen that the conical condenser provides higher heat transfer than the cylindrical condenser. The results of the numerical study indicate that the geometry of the condenser is a main parameter affecting the heat transfer coefficient.

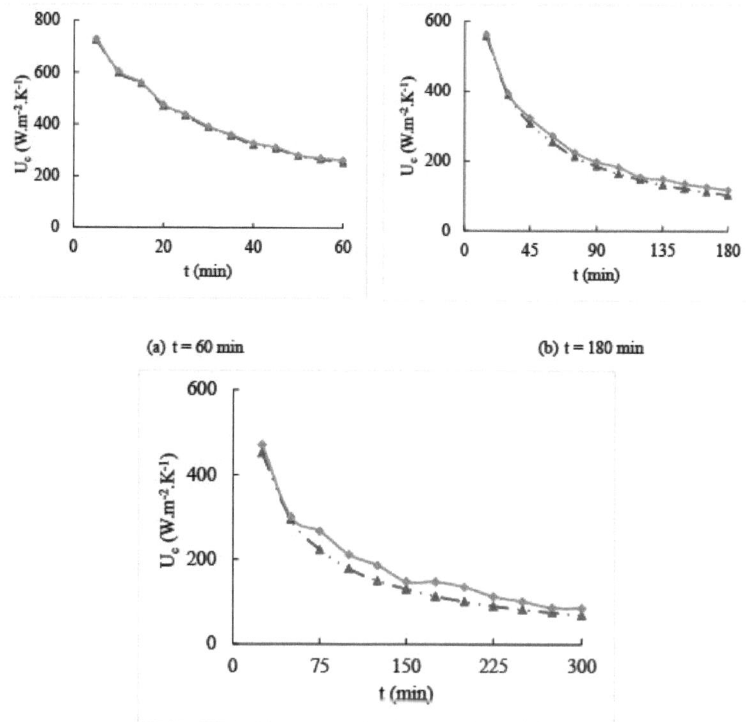

(c) t = 300 min Figure 4. 32. Variation of heat transfer coefficient as a function of condenser geometry

2.3.2.4 Speed field

Figure 4.33 shows the distribution of water velocity in the tank for different heating times. To examine the influence of this parameter, we studied two condensers with two different geometries: a simple cylindrical shape and a conical shape. From the results obtained, it should be noted that the conical condenser leads to an increase in speed for all three heating times compared with the conical condenser. This is due to the increase in water velocity and heat flux provided by the conical condenser in the lower part of the tank. The comparison of the water velocity fields highlights the influence of the condenser geometry on the thermal performance of the domestic refrigerator coupled with a water heater.

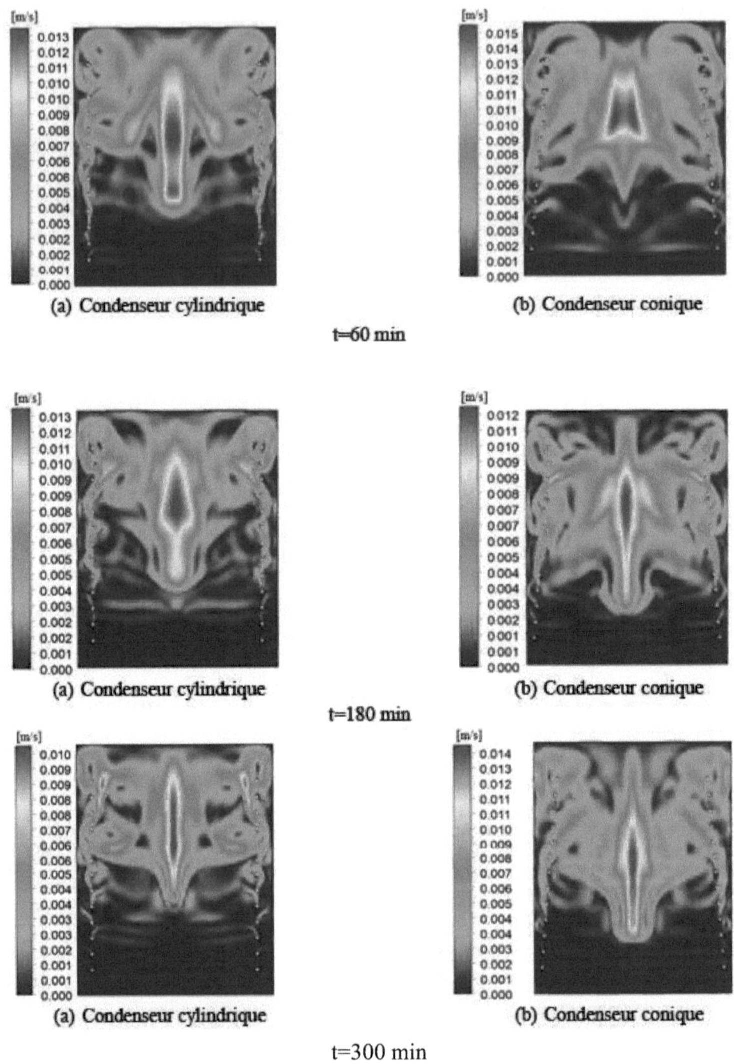

t=300 min

Figure 4. 33. Speed distribution

2.3.2.5 Temperature field

The impact of the condenser geometry on the variation in water temperature is illustrated in Figure 4.34. According to the results obtained, the stratification phenomenon is clear in the tank. After the first heating period t=60 min, it can be seen that there are two parts, with a homogeneous upper part appearing at a temperature of 26°C. On the other hand, the lower part heats up to 20.73°C compared with the initial temperature. The results also show that as the heating time increases from t=60 min to t=300 min, the temperature rises. For the same heating period, the increase in water temperature is significant for a conical condenser. It is therefore clear to see that the conical condenser can considerably improve temperature stratification in the tank. This is mainly due to the heat transfer provided by the larger surface area of the condenser in the lower part of the tank compared with that in the upper part.

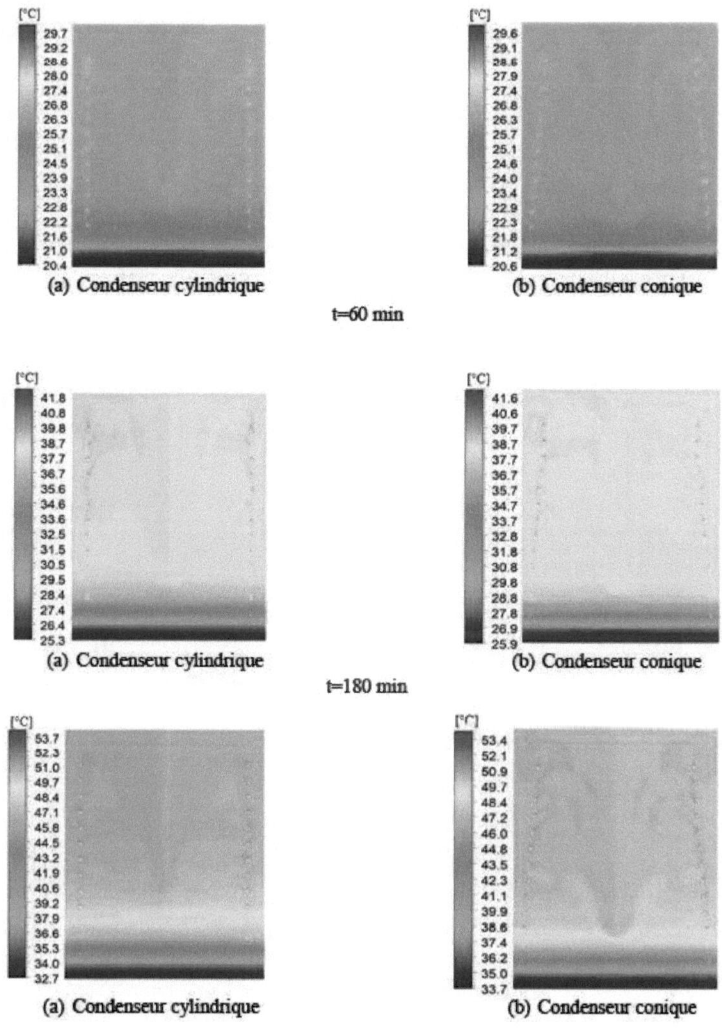

t=300 min

Figure 4. 34. Temperature distribution

2.3.2.6 Pressure field

Figure 4.35 shows the pressure distribution for a cylindrical condenser and a conical condenser for different heating times. As in the case of temperatures, the increase in pressure is significant for a conical condenser.

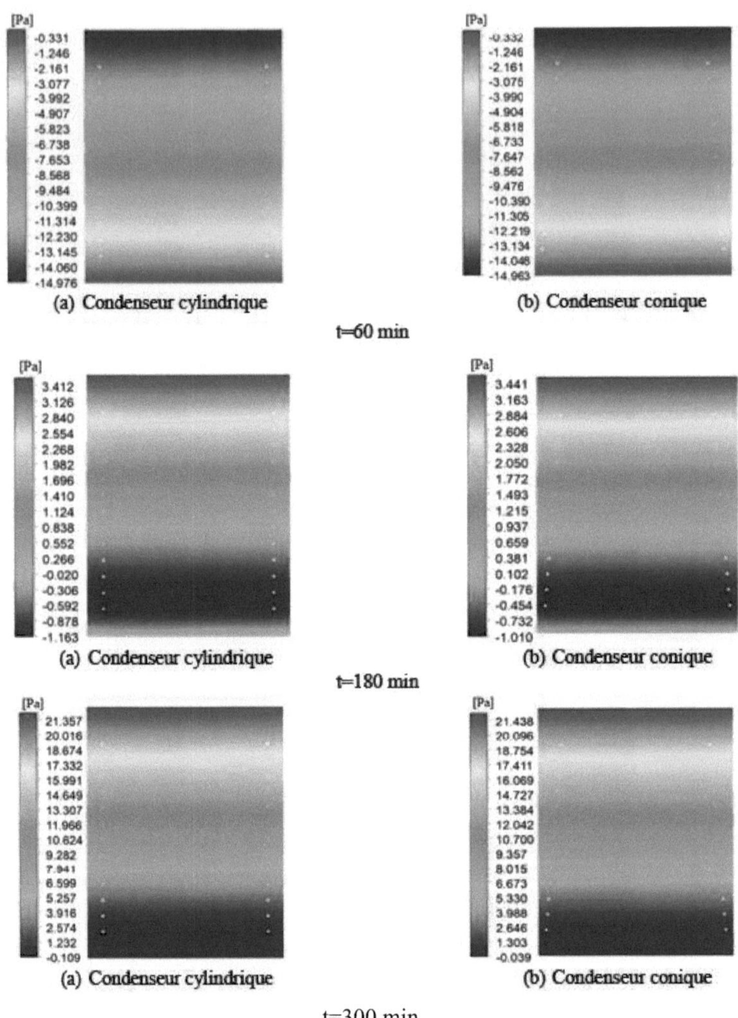

t=300 min

Figure 4. 35. Pressure contours

2.3.3 Performance comparison between conical condenser, constant diameter condenser and variable pitch condenser

2.3.3.1 Temperature profile

Figure 4.36 shows the variation in water temperature with different condenser geometries, namely the constant diameter condenser, the conical condenser and the variable pitch helical condenser. In this case, the turns are very close together in the lower part of the condenser. From the results obtained, it should be noted that the temperature profiles follow the same trend for t=60 min, t=180 min and t=300 min. From this numerical study, it can be seen that there is one main factor affecting the variation of the water temperature in the tank during the three different heating periods. As a result, stratification is reduced and the temperature in the tank is homogenised. Moreover, the impact is even more pronounced for longer heating times. These results show that the temperature distribution is more important for the helicoidal condenser with variable pitch.

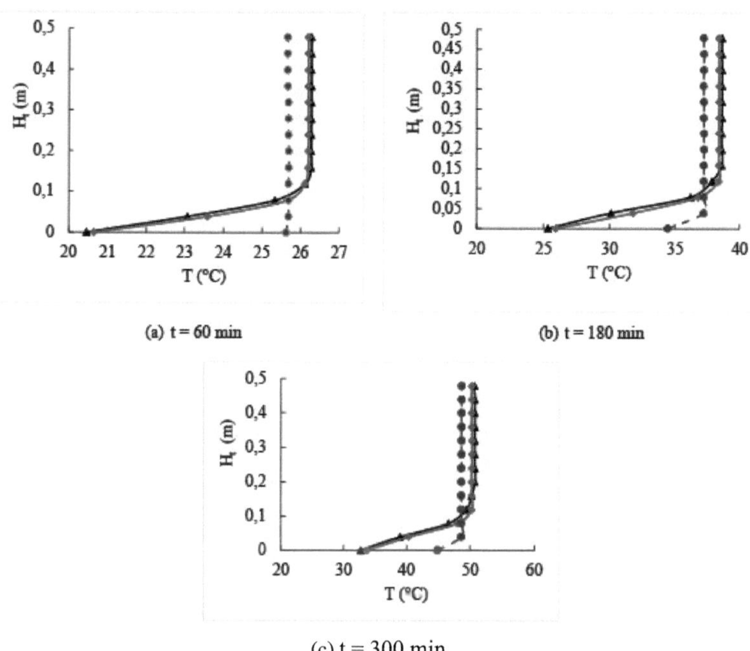

(c) t = 300 min

Figure 4. 36. Variation in water temperature as a function of condenser geometry

2.3.3.2 Heat transfer coefficient profile

The variation of the heat transfer coefficient as a function of the condenser geometry for different heating times is shown in Figure 4.37. From these results, it can be seen that the variable-pitch helicoidal condenser produces a higher heat flux than the two geometries for the same heating period. In fact, the small distance between the condenser turns in the lower part of the tank has a direct effect on these results. This helps us to better design a helical condenser ensure the best heat transfer between the condenser wall and the fluid to be heated.

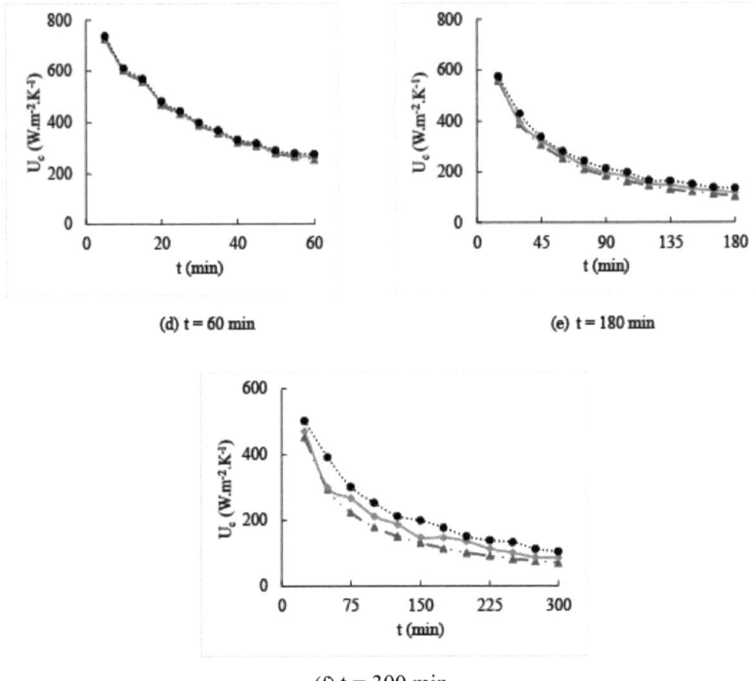

(f) t = 300 min

Figure 4. 37. Variation in heat transfer coefficient as a function of condenser geometry condenser

2.3.3.3 Speed field

In an attempt to better control the geometry, new shapes have been introduced for helicoidal, conical and variable pitch helicoidal condensers. Figure 4.38 shows the variation in water velocity as a function of geometry, with the aim of demonstrating the influence of this parameter on the thermal behaviour of a domestic refrigerator coupled with a water heater. The results obtained show that for t=60 min, the water velocity increases and reaches a maximum value of 0.016 m/s when a variable-pitch condenser is used. As the heating time increases from t=60 min to t=300 min, the velocity increases and the difference between the results becomes clearer. However, with the use of a variable pitch helicoidal condenser, we see an evolution approximately 2 times faster than the other two geometries. It is therefore interesting to see the effect of this parameter on the variation in speed.

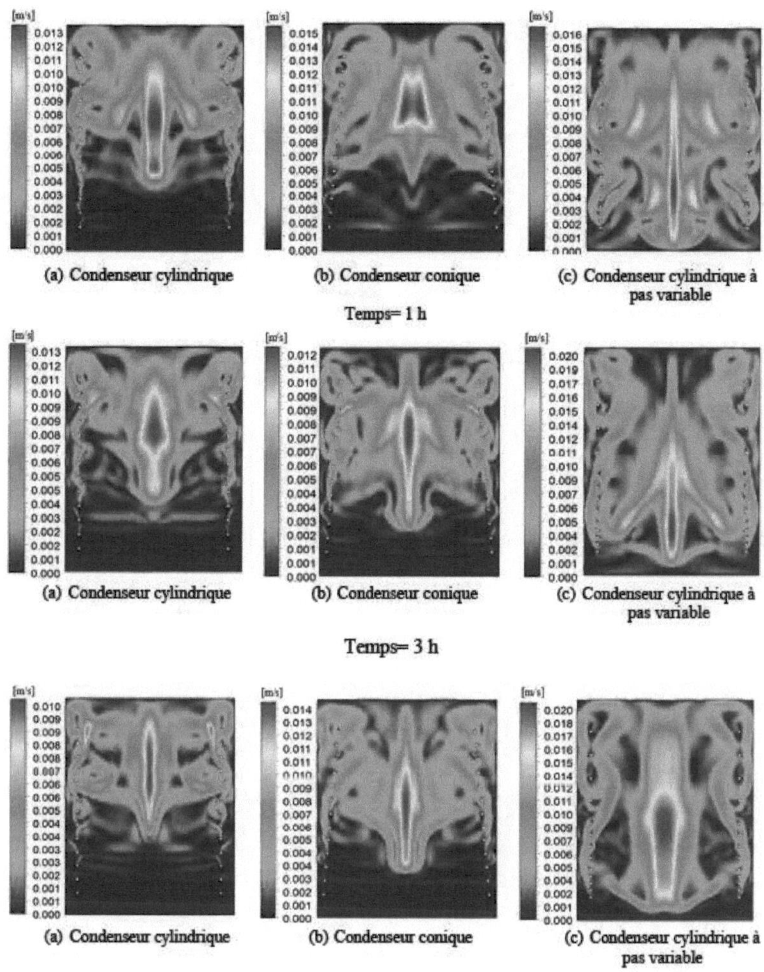

Temps= 5 h

Figure 4. 38. Speed distribution

2.3.3.4 Temperature field

In order to generalise the study to all condenser geometries, we carried out a numerical study on a condenser fully immersed in a 50 litre water tank. For the same operating conditions, we presented the water temperature distribution for different condenser geometries as shown in Figure 4.39. With the change in geometry, the results obtained show a very significant increase in temperature in the lower part of the tank, which leads to a reduction in the stratification phenomenon. For a helical condenser with variable pitch, the maximum temperature and minimum stratification in the tank are reached. This is due to the presence of a variable-pitch condenser in a water tank. Under these conditions, the heat transfer between the cold fluid and the hot wall shows a clear improvement .

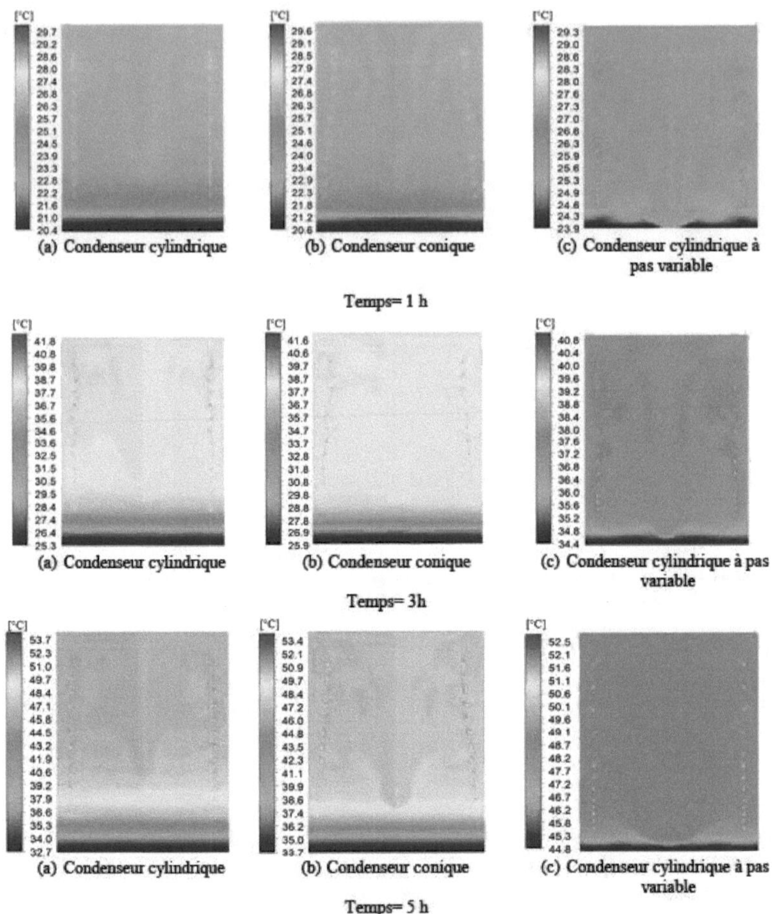

Figure 4. 39. Temperature distribution

2.3 Performance analysis of the domestic refrigerator for different geometries Figure 4.40 shows the variation in the coefficient of performance of a refrigerator with a helical condenser under two types of geometry defined by a variable-pitch helical condenser and a constant-pitch helical condenser. The results show that the performance of a domestic refrigerator coupled with a domestic hot water production unit decreases heating time. This is due to the increase in water temperature in the storage tank, which has contributed to the increase in electrical energy consumption by the compressor. Furthermore, it is easy to see that the performance of the refrigerator with a constant-pitch condenser and a variable-pitch condenser is 4.8 and 5 respectively at the start of operation and 1.6 and 2 respectively after 300 minutes of operation. In addition, these results indicate that the performance decreases with increasing water temperature. It should also be mentioned that the efficiency of the compressor will be affected. When the water temperature in the tank is too high, the heat loss increases, the COP decreases and the normal operation of the compressor is affected. From these results, we can see that the average coefficients of performance for the constant-pitch condenser and the variable-pitch condenser are 3.03 and 3.52 respectively. Consequently, the average coefficient of performance is improved by 16.17% with the use of a variable-pitch helicoidal condenser. As a result of this improvement, the convective heat transfer coefficient increases with the use of a variable-pitch condenser. As a result, the new configuration with a variable-pitch helical condenser leads to faster heat transfer in the lower part of the tank, resulting in improved COP. As already mentioned, the variable-pitch condenser has a good heat transfer distribution in the water, which makes the COP of the refrigerator with a variable-pitch condenser higher than that of a constant-pitch condenser. Thus, it can be confirmed that the performance of the coupled system is significantly improved with the use of a variable-pitch condenser.

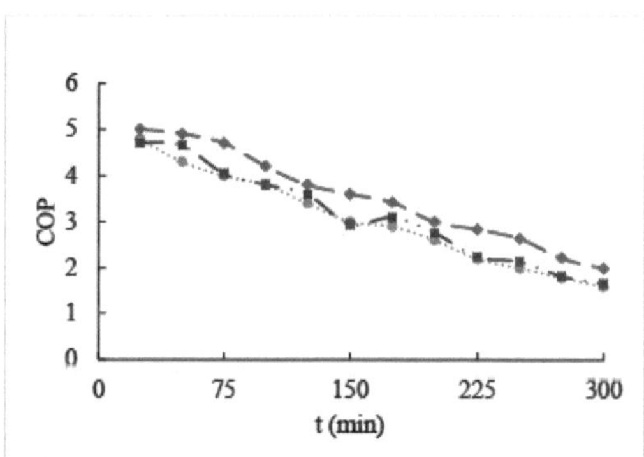

Figure 4. 40. Evolution of the coefficient of performance over time

3 Water desalination part

In this section, we experimentally study an air conditioner for the distillation of brackish water. This study presents the experimental results for each device, such as the temperature profile of the various components of the air conditioner and the production of distilled water over time. A system performance study is then developed for an air conditioner coupled with a water desalination unit. The objective of this section is to experimentally evaluate the performance of the air conditioner coupled with a water distillation unit.

3.1 Distiller

Experimental tests for an air conditioner coupled with a brackish water distillation unit were carried out on 06 March 2018.

3.2 Environmental conditions

The experimental study was carried out in March and April under different ambient conditions for the same geometric parameter of the air conditioner and distiller. The ambient temperature is around 20°C to 21°C.

3.3 Temperature profile

The variation in the temperature of the water in the heating bath and the temperature of the evaporator is illustrated in Figure 4.41. Thanks to the high heat flow given off by the air conditioner condenser, the temperature of the bath and that of the emerging water heating up in the bath increase. In addition, the water temperature increases until it reaches 57°C between 8am and 10am. Under these conditions, the maximum temperature of the glass is around 34°C. This value is lower than that of the water, which allows the water vapour to condense. The results also show that the water temperature changes proportionally with increasing heating time. Although the water temperature and the flow rate recovered increase, the evaporator temperature is not affected and varies between 19°C and 20.5°C.

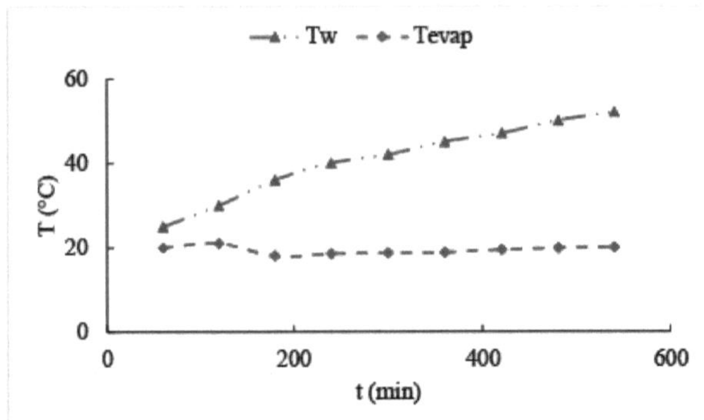

Figure 4. 41. Evolution of water and evaporator temperature over time

3.4 Recovered water flow rate

The hourly flow rate of water recovered during our experimental study is shown in Figure 4.42. According to the results obtained, a remarkable increase in production begins at 4 h. The distilled water is recovered following condensation of the water vapour at the glass. It can also be seen that the variation in distilled water production is due to several factors, such as the heat flow, the ambient temperature, the volume of water and the thickness of the brine. Furthermore, this result indicates that the maximum production is 180 ml/h for the day of 06 March 2019. In fact, increasing the mass of bath water under similar environmental conditions reduces the water temperature, and drinking water production decreases.

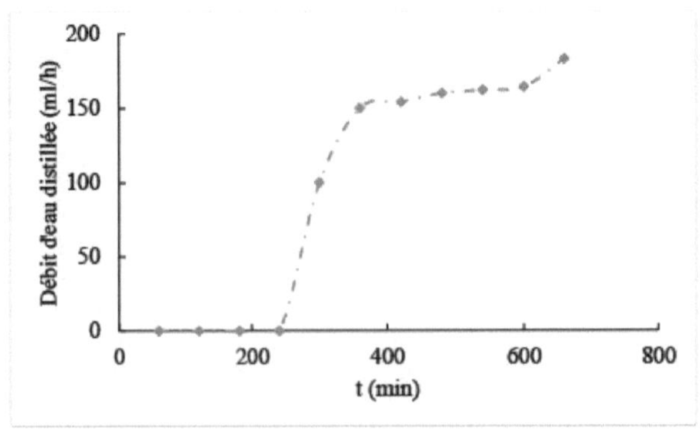

Figure 4. 42. Hourly flow rate of water recovered

3.5 Air conditioner performance

As shown in Figure 4.43, the maximum overall coefficient of performance for an air conditioner coupled to a brackish water distillation unit is of the order of COP = 6.11. From the results obtained, it can be seen that the coefficient of performance increases until it reaches 6 between 11 am and 1 pm. It then decreases as heat transfer between the condenser wall and the water decreases. The drop in COP after a period of operation is mainly due to the increase in the temperature of the water in the bath.

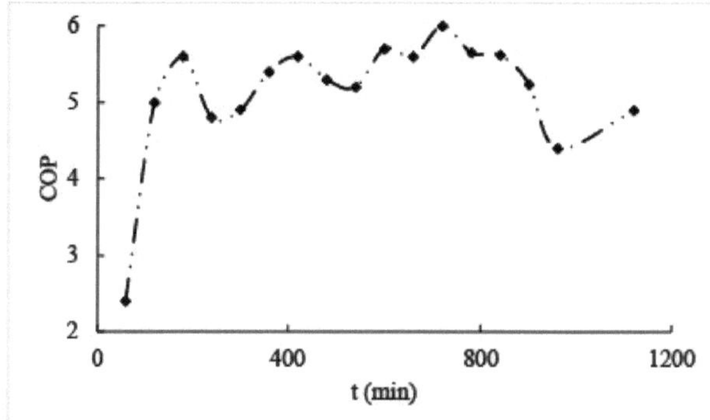

Figure 4. 43. Evolution of the coefficient of performance over time

4 Conclusion

This chapter brings together the various theoretical and experimental results obtained during the analysis of a domestic refrigerator coupled with a water heater and an air conditioner that distils brackish water. In order to achieve our objectives, we have structured the chapter into two parts, which represent the main thrusts of this research.

In the first part, an analysis of the thermal behaviour of a refrigerator coupled to a water heater was presented. We illustrated the various theoretical results of the mathematical models developed to assess the performance of a domestic refrigerator coupled with a domestic hot water production unit. In fact, a parametric study was carried out to improve the thermal behaviour of the helicoï'dal condenser fully immersed in a water tank. We carried out a qualitative analysis of several of the condenser's design parameters, namely tube diameter, condenser pitch and number of turns. The numerical simulation results obtained showed the distribution of velocity, pressure and temperature. The profiles of the heat transfer coefficients, temperature and velocity were also plotted in order to highlight the impact of the condenser's external geometry on its thermal behaviour.

In the second part, we studied the air conditioner used to distil the water. The experiments carried out show that the production of drinking water depends on several parameters, such as the ambient temperature, the operating time and the volume of water. The experiments also show that the production of drinking water is proportional to the temperature of the water and tends to increase as the operating time increases.

General conclusions and outlook

The production of domestic hot water and the desalination of brackish water are fundamental processes that increase electricity in the residential sector and lead to an increase in greenhouse gas emissions. As a result, conventional of meeting these needs are considered to be highly energy-intensive. With the demand to reduce CO_2 emissions and fossil energy consumption, a refrigeration system coupled with a refrigeration machine-heating and/or water desalination unit is seen as a promising and economical solution for reducing electricity and carbon dioxide emissions. This explains the use of coupled refrigeration machines, which have undergone remarkable development in recent years.

In this context, the aim of this thesis is to study a refrigerator that domestic water and an air conditioner that distils brackish water. The aim is to gain a better understanding of the phenomenon of heat transfer by exploiting the thermal rejection of the coupled refrigeration machines in order optimise the quantity of water to be heated on a daily basis. The study must therefore continue, focusing on modelling the operation and coupling of the refrigerator to the water heater, in order to contribute to the development of a domestic water heating unit that is economical, efficient and has no harmful impact on the environment.

In this work, a model describing the operation and coupling of the refrigerator to the water heater was established using the finite volume method. The systems of differential equations forming this model were solved using ANSYS Fluent software. The calculation of heat transfer by convection is carried out in a two-dimensional (2D) Axisymmetric domain with a refined mesh near the condenser tube and the tank wall. The results obtained by this numerical model were compared with those obtained experimentally. Once the mathematical model had been validated, it was published in an internationally-renowned journal.

The results obtained for the various study parameters and for the two refrigerators and the air conditioner were presented. These results also show that :

- The domestic refrigerator can be used to produce hot water without altering its main role in the cooling process.
- As the water temperature rises, the convective exchange coefficient decreases, reaching a minimum value of 68.64 $W.m^{-2}.K^{-1}$ after 300 min of heating.
- The hot water flows continuously upwards and accumulates in the upper part of the tank to form a high-temperature zone. We have also observed that the temperature of the water in the upper part of the tank is obviously higher than in the lower part.
- The distance between the condenser coils is an important factor influencing the thermal performance of a refrigeration machine used to produce domestic hot water.
- The geometry of the condenser is the main parameter affecting the temperature of the water inside the tank.
- Increasing the condenser pitch reduces the heat transfer coefficient between the tube wall and the water. As a result, heat loss increases.
- The diameter of the condenser tube is the main parameter affecting water velocity.
- The diameter of the tube is a major factor affecting the thermal performance of a domestic fridge coupled with a water heater.
- Reducing the number of turns and at the same time increasing the diameter of the condenser also ensures good turbulence in the upper part of the tank and therefore good heat transfer by natural convection.

■ The use of a helical condenser with N=11 turns is therefore beneficial for improving heat exchange between the condenser walls and the water.

■ Comparison of the results confirms that the external shape of the helicoidal condenser has a direct effect on the distribution of water velocity.

■ The variable-pitch condenser improves the convective transfer coefficient between the water in the receiver and the refrigerant in the condenser.

■ The variation in the static pressure of the water is proportional to the geometry of the condenser.

■ The use of a cone-shaped condenser to heat the water therefore proves to be the ideal solution.
beneficial for improving the heat transfer phenomenon between the tube wall and the water outside.

One of the aims of this thesis is to work on improving the thermal performance of refrigeration machines used to produce domestic hot water using the heat released by the condenser. In particular, we propose to modify the geometric shapes of the condenser and to couple a variable-pitch condenser with a variable-diameter condenser.

References

1- Carolina Flores Bahamonde, Étude des transferts de masse et de chaleur au sein d'un absorbeur eau/bromure de lithium, Thesis 7 August 2006

2- Oussama Ibrahim et al, Air source heat pump water heater: Dynamic modeling, optimal energy management and mini-tubes condensers, Energy (2013), 2013.11.017.

3- Récupération de chaleur et utilisation des rejets thermiques, Planification, construction et exploitation rationnelle de la récupération de chaleur et de l'utilisation des rejets thermiques, RAVEL dans le domaine de la chaleur, Cahier 2.

4- Hong Li et al, Study on performance of solar assisted air source heat pump systems for hot water production in Hong Kong, Applied Energy 87 (2010) 2818-2825.

5- Jiaheng Chen, Jianlin Yu, Theoretical analysis on a new direct expansion solar assisted ejector-compression heat pump cycle for water heater, Solar Energy 142 (2017) 299307.

6- Xinhui Zhao et al ,Experimental Study on Heating Performance of Air - source Heat Pump with Water Tank for Thermal Energy Storage, Procedia Engineering 205 (2017) 2055-2062.

7- Xiaolin Sun et al, Performance comparison of direct expansion solar-assisted heat pump and conventional air source heat pump for domestic hot water, Energy Procedia 70 (2015) 394 - 401.

8- Nannan Dai, Shuhong Li, Simulation and performance analysis on condenser coil in household heat pump water heater, Sustainable Cities and Society 36 (2018) 176-184

9- Qiang Ye, Shuhong Li, Investigation on the performance and optimization of heat pump water heater with wrap-around condenser coil, International Journal of Heat and Mass Transfer 143 (2019) 118556

10- Andreas Genkinger et al, Combining heat pumps with solar energy for domestic hot water production, Energy Procedia 30 (2012) 101 - 105.

11- Mustafa S. Mahdi, Hameed B. Mahood, Alasdair N. Campbell, Anees A. Khadom, Experimental Study on the Melting Behavior of a Phase Change Material in a Conical Coil Latent Heat Thermal Energy Storage Unit, Applied Thermal Engineering (2019), doi: https://doi.org/10.1016/j.applthermaleng.2019.114684

12- V. Skrivan, Utilisation of the heat of condensation from medium capacity refrigerating units for heating water, International Journal of Refrigeration Volume 7 Number 1 January 1984.

13- D. Carbonell et al, Potential benefit of combining heat pumps with solar thermal for heating and domestic hot water preparation, Energy Procedia 57 (2014) 2656 - 2665.

14- Flora B.F et al, Heat Pump for Heating Water for Domestic Purposes Using a Varying Speed Compressor Control.

15- Hong Li et al, Study on performance of solar assisted air source heat pump systems for hot water production in Hong Kong, Applied Energy 87 (2010) 2818-2825.

16- Mei VC et al, A study of a natural convection immersed condenser heat pump water heater. ASHRAE Trans 2003; 109 (2):3-8.

17- Jianbo Qin et al, Experimental investigation of gas bubble diameter distribution in a domestic heat pump water heating system, Energy Procedia 123 (2017) 361-368.

18- Romdhane Ben Slama, Thermodynamic heat water by the condenser of refrigerator, Int. Symp. on Convective Heat and Mass Transfer in Sustainable Energy April 26 - May 1, 2009, Tunisia.

19- Walid Youssef et al, Effects of latent heat storage and controls on stability and performanceof a solar assisted heat pump system for domestic hot water production, Solar Energy 150 (2017) 394-407.

20- Tianji Liu et al, Experiments of a Heat Pump Water Heating System Using Stored Solar Energy to Defrost, The 8th International Conference on Applied Energy - ICAE2016, Energy Procedia 105 (2017) 1130 - 1135.

21- Kadir Bakirci et al, Experimental thermal performance of a solar source heat-pump system for residential heating in cold climate region, Applied Thermal Engineering 31 (2011) 1508-1518.

22- Ralf Dott et al, System evaluation of combined solar & heat pump systems, Energy Procedia 30 (2012) 562 - 570.

23- Sara Eicher et al, Solar assisted heat pump for domestic hot water production, Energy Procedia 30 (2012) 571 - 579.

24- Romdhane Ben Slama, Refrigerator Coupling to a Water-Heater and Heating Floor to Save Energy and to Reduce Carbon Emissions, Computational Water, Energy, and Environmental Engineering, 2013, 2, 21-29.

25- Jing-Wei Peng et al, Performance comparison of air-source heat pump water heater with different expansion devices, Applied Thermal Engineering 99 (2016) 1190-1200.

26- G.G. Momin et al, Cop Enhancement of Domestic Refrigerator by Recovering Heat from the Condenser, International Journal of Research in Advent Technology, Vol.2, No.5, May 2014 E-ISSN: 2321-9637.

27- Lakshya Soni et al, Waste heat recovery system from domestic refrigerator for water and air heating, International Journal of Engineering Sciences and Research Technology.

28- Jadhav. P. J et al, Heat Recovery from Refrigerator Using Water Heater and Hot Box, International Journal of Engineering Research & Technology (IJERT), ISSN: 22780181, Vol. 3 Issue 5, May - 2014

29- N. B. Chaudhari et al, Heat Recovery System from the Condenser of a Refrigerator - an Experimental Analysis, ISSN (Print): 2319-3182, Volume -4, Issue-2, 2015.

30- Sreejith K. et al, Experimental Investigation of a Household Refrigerator using Aircooled and Water-cooled Condenser, Research Inventy: International Journal of Engineering And Science, Vol.4, Issue 6 (June 2014), PP 13-17,ISSN: 2278-4721, Issn (p):2319-6483, www.researchinventy.com.

31- Omkar Borkar et al, A Review on Utilization of Waste Heat from a Refrigerator, IJSRD - International Journal for Scientific Research & Development| Vol. 5, Issue 01, 2017 | ISSN (online): 2321-0613.

32- Prashant.S.Pathak et al, Review Study of Waste Heat Recovery using Refrigeration System, GRD Journals- Global Research and Development Journal for Engineering | Volume 2 | Issue 5 | April 2017, ISSN: 2455-5703.

33- Pratik Kumbhar et al, Design and Development Waste Heat Recovery from Domestic Refrigerator, International Journal of Advance Research, Ideas and Innovations in Technology, ISSN: 2454-132X, Volume3, Issue2. Available online at www.ijariit.com.

34- Soma A. Biswas et al, Waste heat recovery from domestic refrigerator , International Journal of Current Engineering and Scientific Research (IJCESR), ISSN (PRINT): 2393-8374, (ONLINE): 2394-0697, Volume 4, Issue9, 2017.

35- Sreejith K., Experimental Investigation of A Domestic Refrigerator Having Water-Cooled Condenser Using Various Compressor Oils, International Journal Of Engineering And

Science, 2278-4721, Vol. 2, Issue 5 (February 2013), Pp 27-31, Www.Researchinventy.Com.

36- Sreejith K et al, Experimental Investigation of a Household Refrigerator Using Evaporative-Cooled Condenser, International Journal of Engineering And Science, Vol.7, Issue 4 (April 2017), PP -01-06, 2278-4721, Issn (p):2319-6483, www.researchinventy.com.

37- Tanmay Patil et al, A Review On Recovering Waste Heat From Condenser Of Domestic Refrigerator, International Journal of scientific research and management (IJSRM), Volume||3|Issue||3|Pages|| 2409-2414||2015||, Website: www.ijsrm.in ISSN: 2321-3418.

38- Tarang Agarwal et al, Cost-Effective COP Enhancement of a Domestic Air Cooled Refrigerator using R-134a Refrigerant, International Journal of Emerging Technology and Advanced Engineering, Website: www.ijetae.com ISSN 2250-2459, ISO 9001:2008 Certified Journal, Volume 4, Issue 11, November 2014.

39- Hanning Li et al, Techno-economic feasibility of absorption heat pumps using wastewater as the heating source for desalination, Desalination 281 (2011) 118-127.

40- Farzaneh Mahmoudi et al, Sustainable seawater desalination by permeate gap membrane distillation technology, Energy Procedia 110 (2017) 346 - 351.

41- Giuseppe Franchini et al, Modeling of a solar driven HD: HumidificationDehumidification desalination system, Energy Procedia 45 (2014) 588 - 597.

42- Jinzeng Chen et al, A discussion of "Heat pumps as a source of heat energy for desalination of seawater", Desalination 169 (2004) 161-165.

43- S. Jasechko, Z.D. Sharp, J.J. Gibson, S.J. Birks, Y. Yi and P.J. Fawcett, Terrestrial water fluxes dominated by transpiration, Nature 496 (2013), pp. 347-350.

44- R. Tripathi and G.N. Tiwari, Performance evaluation of a solar still by using the concept of solar fractionation, Desalination 169 (2004), pp. 69-80

45- A. Kettab, Les ressources en eau en Algérie: stratégies, enjeux et vision, Desalination 136 (2001), pp. 25-33.

46- A. Maurel, Dessalement de l'eau de Mer et Des Eaux Saumâtres et Autres Procédés Non Conventionnels d'approvisionnement En Eau Douce, Éditions Tec & Doc, 2001.

47- A.A. El-Sebaii and E. El-Bialy, Advanced designs of solar desalination systems: A review, Renewable and Sustainable Energy Reviews 49 (2015), pp. 1198-1212.

48- O. Halloufi, Etude de la performance d'un distillateur solaire par un système de pré-chauffage solaire de l'eau saumatre, (2010), .

49- S. Satcunanathan and H.-P. Hansen, An investigation of some of the parameters involved in solar distillation, Solar Energy 14 (1973), pp. 353-363

50- Sami Missaoui, Zied Driss, Romdhane Ben Slama and Bechir Chaouachi, Experimentally validated model of domestic refrigerator with immersed condenser coil for water heating, International Journal of Air-Conditioning and Refrigeration, https://doi.org/10.1142/S201013252150022X

51- I.P. Holman, Experimental Methods for Engineers, sixth ed. McGraw-Hill, New York, 1994, 48.

52- ANSYS Fluent user's Guide; Release 13.0

53- Salim Ibrahim Hasan et al. Numerical Study of Refrigerant Flow in CapillaryTubeUsingRefrigerant(R134a),https://www.researchgate.net/publication/322 086504. 27 December 2017.

54- Missaoui Sami, Ben Slama Romdhane and Chaouachi Bechir, Optimum length of a condenser for a domestic refrigerator for water heating, Australian Journal of Basic and Applied Sciences, 13(7): 1-5.

55- Sami Missaoui, Zied Driss, Romdhane Ben Slama and Bechir Chaouachi, Numerical analysis of the heat pump water heater with immersed helically coiled tubes, Journal of Energy Storage 39 (2021) 102547
56- Nannan Dai, Shuhong Li, Qiang Ye; "Performance analysis on the charging and discharging process of a household heat pump water heater"; International Journal of Refrigeration 98 (2019) 266-273.
57- Wenzhe Li, PegaHrnjak (2018). Experimentally validated model of heat pump water heater with a water tank in heating-up transients. International Journal of Refrigeration(88), 420-431.
58- D.D. Wang, Research on temperature field and flow field in tank of ASHPWH, Beijing Institute of Civil Engineering and Architecture Doctor of Philosophy, 2006.
59- Jie JI, Huide FU, Hanfeng HE, Gang PEI; "Performance analysis of an air-source heat pump using an immersed water condenser"; Front. Energy Power Eng. China 2010, 4(2): 234-245

Appendix 1

The heat flux variable code as a function of time for the UDF (User Defined Function) is defined as follows:

```
#include "udf.h"
DEFINE_PROFILE (ramp_heat,thread,position)
{
float t,heat;
face_t f;
t=RP_Get_Real ("flow-time");
heat= q(t)i = A.t^3 + Bt^2 + C.t + D ;
begin_f_loop (f,thread)
{
F_PROFILE (f,thread,position) =heat;
}
end_f_loop (f,thread)
}
```

Appendix 2

1. International publications

- Sami Missaoui, Zied Driss, Romdhane Ben Slama and Bechir Chaouachi, Numerical analysis of the heat pump water heater with immersed helically coiled tubes, Journal of Energy Storage 39 (2021) 102547
- Sami Missaoui, Zied Driss, Romdhane Ben Slama and Bechir Chaouachi, Experimentally validated model of domestic refrigerator with immersed condenser coil for water heating, International Journal of Air-Conditioning and Refrigeration, https://doi.org/10.1142/S201013252150022X
- Missaoui Sami, Ben Slama Romdhane and Chaouachi Bechir, Optimum length of a condenser for a domestic refrigerator for water heating, Australian Journal of Basic and Applied Sciences, 13(7): 1-5.
- Sami Missaoui, Zied Driss, Romdhane Ben Slama and Bechir Chaouachi, Experimental and numerical analysis of a helical coil heat exchanger for domestic refrigerator and water heating, International Journal of Refrigeration 2022.
- Sami Missaoui, Zied Driss, Romdhane Ben Slama and Bechir Chaouachi, Effects of pipe turns on vertical helically coiled tube heat exchangers for water heating in a household refrigerator, International Journal of Air-Conditioning and Refrigeration 2022.
- Sami Missaoui, Zied Driss, Romdhane Ben Slama and Bechir Chaouachi, Theoretical Analysis on a Household Heat Pump Water Heater With Immersed Condenser Coil, Chapter Book: Advances in the Modelling of Thermodynamic Systems (pp.236-252).

2. National and international communications

- Missaoui S., Ben Slama R., Chaouachi B. Optimum length of a condenser for domestic refrigerator for water heating, Third International Conference on Advances in Mechanical, Industrial and Mechatronics Engineering (ICAMIME 2019), April 19-20, 2019, Tunis, Tunisia **[Oral presentation].**
- missaoui.S, Ben Slama .R, Chaouachi .B, Utilization of Waste heat from Domestic Refrigerator for Water heating-an Experimental Analysis, The 6th International Conference on Green Energy and Environmental Engineering GEEE-2019, 27 - 29 April 2019 - Tabarka, Tunisia **[Oral presentation].**
- Sami Missaoui , Romdhane Ben Slama , Bechir Chaouachi, Theoretical analysis on a household heat pump water heater with immersed condenser coil, First International Advanced Modeling of Thermodynamic Systems (AMTS2021, Online March,10- 11,2021 ISSAT Gabes **[Oral presentation].**
- Sami Missaoui, Romdhane Ben Slama, Bechir chaouachi , Waste heat recovery from helical condenser coil of domestic refrigerator and applications ,3rd Euro-Mediterranean Conference For Environmental Integration, 10-13 JUNE 2021, Sousse, Tunisia. **[Oral presentation]**
- Sami Missaoui, Romdhane Ben Slama, Béchir Chaouachi, Desalination of seawater using waste heat from condenser of air conditioning, International Meeting on Advanced Technologies in Energy and Electrical Engineering, 28-29 Nov. 2019 Tunis, Tunisia **[Oral presentation]**
- Sami Missaoui, Zied Driss, Romdhane Ben Slama and Bechir Chaouachi, Numerical investigation of an air-source heat pump water heater with immersed helically coiled tube

heat exchanger, International Conference on Mechanics and Energy (ICME'2021), 27-29 December, 2021, Sousse, TUNISIA **[Oral presentation].**

•

Printed by Books on Demand GmbH, Norderstedt / Germany